MEET THE BOYS OF CASPER

ISBN: 978-0-9990017-0-7

Printed in the United States of America

First Printing, 2017

This book is a fictional work. The names, characters, places, and incidents are products of the author's imagination, or the author has used them fictitiously.

Cover design and formatting by Damonza.com

Published by Guy Talk Press
www.guytalkpress.com

GUY TALK PRESS

For My Friends
Thanks for sharing a smile, a moment, or a laugh.

MEET THE BOYS OF CASPER

DALLAS JONES

Synopsis

The almost incessant wind swept over six boys and cried out three questions.

Where are you?

In Casper, a town not so different from many. In Wyoming, a place like no other. In the United States, the greatest country on Earth—at least it used to be.

When are you?

1975.

Who are you?

This question the boys could not answer. They recalled who they used to be, and sometimes they understood who they wanted to be, but today, like the wind, their identities passed through their outstretched fingers.

Growing up ain't always easy.

PRELUDE

Consciousness And Conformity Are Cosmic Ideals
But Bitchyness Is Universel

They read the sign. What did it mean? When did it happen? Who wrote it? Was it a slogan? Was it a warning or perhaps a threat? Whatever it was, it was unnerving. Whatever it was, it proved that changes—most likely unwanted changes—were taking place. Whatever it was, it needed to be addressed immediately.

These questions and comments referred, of course, to line two of the inscription. Line one had been chiseled into the monolith during its creation long ago, and that creed had been revered by generations as their guiding principle. The monolith itself represented all things in a clear and concise manner, but line two appeared from seemingly nowhere and polluted the shrine. This terrible graffiti shook the soul of a civilization.

Although the culprit was currently unknown, much could be surmised. The different font style indicated the author was clearly antiestablishment. The sheer callousness of the act suggested this creature was sinister. Finally, all who viewed the freshly scribbled verse knew beyond a shadow of a doubt the perpetrator couldn't spell worth shit.

In response to the situation, a vessel was launched into space. It carried only one passenger, and during its journey it would have to navigate both time and distance. Databanks had been scoured to determine potential sources for answers. The most likely candidate was identified and a destination course set. Those participating in the search were quite disappointed in the result. The chosen site was a small planet having been briefly studied long ago. It had been deemed trivial back then, but the research team noted irrationality as one of its features.

Soon after the vessel launched, the lone passenger in the spacecraft reflected on his assignment. He knew he would have to conduct his analysis using the language of the native population and operate within the confines of their meager understanding of the universe. He sighed at this realization because in galactic terms they were dumber than space dust. His mission was to study a small population of human beings on a planet they called Earth.

As his first assignment, the traveler randomly selected a finite time period and geographic area for detailed study. Time, as Earthlings knew it, was defined in relation to the life of a benevolent fellow who roamed the planet long ago. The traveler drew four digits from a random number generator. The first digit was *one* followed by *nine*. On the backend, the numbers *seven* and *five* were drawn.

Then he grabbed a giant replica of the United States, closed his eyes, turned himself in a circle three times, and placed a finger on the map. The location under his finger was a state named Wyoming and a region called Natrona County.

THE SELECTION

It rained like a cow peeing on a flat rock.

The space traveler considered the statement with equal parts annoyance and bewilderment. He encountered the sentence while studying a recent edition of the Casper newspaper, a human form of broadcast communication using printed language. Now that he had stabilized his vessel into a fixed orbit around Earth, the traveler's next task was to familiarize himself with the local culture within the geographic region designated for observation. The phrase was contained in a story about a spectacular cloudburst that happened the previous day in the canyon areas of Utah. The reference jumped out as an obvious sign of irrationality.

Why would humans employ the elimination mechanism of a domesticated animal as a precipitation metric, he wondered. Further, the multidirectional spattering of droplets would make reliable collection practically impossible. The traveler was frustrated by this bizarre human methodology because he knew they had mastered basic scientific principles. They had established standardized units of measure such as inches and centimeters. This comparison of atmospheric moisture and animal urine was simply ridiculous—or maybe it wasn't.

Perhaps he was overlooking something. He vaguely remembered instances when a second examination of local customs in other worlds revealed meaningful insight. He was out of practice in the spy game because so much time had passed since his last reconnaissance.

As the traveler rolled his neck to ward off stiffness, he noticed a second human document lying atop the console. Earthlings called this item a dictionary, and it contained all the words in the English language. The traveler had opened the book a short time earlier to pages 574 and 575, and now it recaptured his attention. The word *puffy* was highlighted on the top left-hand corner of page 574, and the word *quail* was listed on the top right-hand corner of page 575.

A disappointed feeling permeated through him as he examined the words on these pages. Actually, the traveler was bothered by one particular word. He stewed in frustration because no rational reason could account for his reaction; it was simply a word. Yet, he couldn't shake this sensation.

Out of curiosity, the traveler had decided to investigate whether the unique designation assigned to him just before he departed from his home planet carried any meaning on Earth. He had been given the four-character identifier, PUUK. During his review of human communication techniques, he learned the dictionary presented all the words in alphabetical order, and as he skimmed through the words on page 575, no entry for PUUK appeared. The traveler then wondered if some phonically similar word might exist. When he pored through the entries on page 574, he found a potential match. The English word was *puke*. The traveler leaned forward to read the associated definition. An instant later, his shoulders drooped; his head fell.

PUUK's large black eyes blinked a few times while he processed his reaction. *It's just a word and a coincidence*, he thought. *There's no connection*.

The traveler regrouped his thoughts and turned his attention back to the newspaper article with the mysterious simile. His directives for information gathering included terrestrial visits, and PUUK was anxious to see, hear, and touch this new world.

Every mission must commence somewhere, he mused. An investigation into this irrational measurement technique qualified as a reasonable place to start.

To reach the vessel's surveillance cameras, PUUK walked across the control room. He adjusted the equipment, reduced the size of the area being viewed, and then shifted the focus from the city of Casper to a non-populated area directly to the east. PUUK zoomed in a second time making finer landscape details visible. He guided the camera controls back and forth as he scanned for a herd of cattle. The traveler wanted to locate a group of animals away from any human population. He didn't want to be seen by his unknowing hosts, not yet anyway.

After a few sweeps, PUUK spotted an ideal location. He counted thirty-four cows grazing in open grassland with a large rock outcropping nearby. Throughout the rocky area, a pine forest flourished. The setting provided beneficial cover from which he could observe the animals.

PUUK readied his terrestrial kit and checked it twice to ensure all the essential items were included. He then entered the transport bay and jettisoned toward Earth. PUUK's method for shooting down to the planet's surface involved technology currently incomprehensible to humans.

Basically, PUUK sat in a foam-like seat with a high back but no legs. Its base was rounded, and PUUK could rock the chair back and forth using his body. In his hand, the alien operated some sort of control device whose size resembled a little black book. Upon the traveler's instruction, an outer shell formed around PUUK and his belongings. From a human's standpoint, the shell could best be described as a kind of plastic wrap with some amazing qualities. The shell contoured around its inner contents allowing for a sufficient pocket of air to create space inside the vessel. After launching, the shell reformed into an optimal shape for entering Earth's atmosphere. Finally, upon landing the shell and the foam seat decomposed rapidly leaving only small traces of a dusty residue. His return trip would again use another plastic wrap and a propulsion device that converted simple hydrogen into energy.

PUUK's impressive expertise terminated once he landed. His arrival was undetected by humans but caused quite a stir among the cattle. As he scrambled to move from the edge of the pasture and into the rocks for cover, his awkward gait caught the attention of the animals nearest him. They misinterpreted the newcomer as a deliverer of feed and followed him toward the rocks.

The traveler's heart beat faster. He didn't want to harm the creatures, but they were bringing unwanted attention to his whereabouts. PUUK mentally sorted through all the human films he had watched during his journey to determine a course of action. Unfortunately for him, *Rawhide* hadn't been part of the package. He tried to recall any examples of interactions between humans and animals, but only a sketchy memory of some man in a suit babbling about talking to the animals came forth. Instead of wildly swinging his arms in a head-'em-up, move-'em-out sort of motion, PUUK knelt and clasped his hands into a single fist. He began chanting.

"Disperse. Disperse."

The cows stood their ground completely unimpressed. After waiting patiently to receive some type of food reward, one cow gave up. She mooed and turned back toward the grassy pasture. The others followed.

Now that he was rid of the cattle, PUUK turned his large, bald head back and forth searching for any indications that he'd been discovered. Sensing none, the alien leaned against the edge of a boulder and relaxed his breathing pace. As calmness set in, PUUK became cognizant of the atmospheric conditions surrounding him. Sol's energy radiated at an almost ideal temperature, and he detected only a slight movement in the air. The space traveler's uninterrupted peace lasted only a few moments. A deer fly, complements of the cows' visit, bit PUUK on the arm.

"Wa!" he cried out. *Wa* was a term used on his home planet to express surprise. PUUK couldn't remember the last time he'd uttered that term, but he shouted it a second time when another fly munched on his neck. Alarmed by his own shout, PUUK again scanned his immediate surroundings to ensure he was alone. With no signs to the contrary, he retrieved a small canister from his terrestrial kit and sprayed a fine mist onto his body. The flies would cause no more wa's on this day.

The bellowing of a nearby cow reminded PUUK of his trip's purpose. He watched the animals and waited for the desired event. After several go rounds of tilting his head to gain a better view of a selected animal's underside, PUUK began to feel an ache in his back and realized his efforts were futile. He admonished himself for acting spontaneously.

Too many variables, he thought. *First, cows don't pee on demand, and second, the likelihood of a flat rock lying beneath them when they do pee is entirely random.* He felt comfortable with this latter conclusion because he had witnessed a few cases of urination, but at no time did a cow seek out a flat rock. As for the first observation, well, PUUK believed it to be true after he again knelt and clasped his hands while whispering "Wizz, wizz," to no avail.

Still, PUUK deemed his first trip to the area humans called Wyoming a success. Except for the flies, he encountered some delightful sensations. The fresh fragrance of pine trees mixed with the distinct scent of sagebrush and created a complex and dazzling smell. The panorama of buckled grassland with the small, blue mountain in the distance sedated his anxieties. He was thrilled to have escaped the confines of his spacecraft; his interstellar imprisonment had lasted too long.

The sensations from his new surroundings almost lulled PUUK into a delicate snooze, but any chance of a nap ended when a human voice startled the alien.

"Hey, Rankin! Come over here. I think I've found an arrowhead."

PUUK coiled into a ball behind the boulder and held his breath. His mind raced through a list of possible counteractions to take if he'd been spotted.

"C'mon, man! Hurry up," the same voice repeated.

"Hold your horses. I gotta tie my shoe," yelled a second, unseen player.

PUUK remained still and listened carefully for any oncoming encroachment into his immediate area. There was none.

When the second voice spoke again, it projected a softer, conversational tone and seemed more distant than it initially sounded. Also, the alien didn't hear any accompanying footsteps, so PUUK surmised the humans were farther away from him than he originally feared. Was it a safe distance? He wasn't certain.

"Okay, show me what you've found."

"Look, it's got a pointed end, and there are some little notches where it was tied to the arrow."

After a pause, the second voice replied. "Yeah, it's an arrowhead, all right; a nice one, too. You can see some of the divots where it's been chiseled."

PUUK peeked over the boulder and spotted the humans immediately. In human distance, they stood roughly one hundred feet from him. They were facing toward the mountain and not directly looking his way. By their size, the space traveler judged them to be *transitioners*, those beings existing in the state of flux where they morph into their adult form. *Humans have another name for it*, PUUK thought. *What are they called?*

He processed his inventory of English vocabulary as he continued to watch the humans in the field. Then he remembered. *Teenagers. Yes, they're called teenagers.*

Mentally, PUUK recorded many other characteristics about these teenagers. They were male and similar in height to each other and to him. However, they were much thicker than him in both their trunks and limbs. They wore some sort of blue fabric on the lower half of their body and a garment with a rounded neckline and only partial covering for their arms on top. One teenager wore a white garment while the other male was dressed in red. PUUK rocked back unexpectedly when some type of metallic shine glimmered from the mouth of the teenager in white.

"What're you two doing, comparing the size of your wieners?"

A third figure appeared in PUUK's field of vision. Another male teenager, who was even more robust in size than the original two, walked toward his companions. A stand of pine trees had successfully blocked PUUK's view of the newcomer until just now.

"We were waiting for you to unzip our pants and measure," replied the boy with the metallic mouth.

"Borias found an arrowhead, a good one, too," added the red-shirted teenager.

PUUK noticed the new human was carrying some type of partially filled bag. When he looked back at the other two, he spotted similar containers lying on the ground near them.

"An arrowhead? Let me see."

The alien watched as the teenager in white—didn't his companion just identify him as Borias—displayed a small item to the third human.

"Well, kiss my ass and call me Mabel, that's pretty cool," said the third boy.

"Did you find anything, Randy?" asked the boy in the red shirt.

PUUK again picked up on the identifier. *Randy*, he said to himself.

"Oh, yeah, I found something, but it's a piece of shit."

Borias and the other teenager said nothing but waited expectantly.

"No, seriously. It's a piece of shit, actually several pieces of shit. I found rabbit turds, a cow pie, and some horse poop. Instead of a rock collection, I'm gonna turn in a bunch of dried animal shit and tell Fickel to separate them into igneous, sedimentary, and metamorphic."

"Hey, that's good," agreed Borias. "Then you can ask Fickel to arrange them according to Mohs scale of mineral hardness. Can't you just see him grinding a cow pie against a piece of quartz, and it crumbles into his hand. He'd be oblivious, of course, only concentrating on whether it deserved a score of one or two."

The red-shirted human spoke. "Guys, that doesn't make any sense. Poop couldn't possibly be considered a mineral unless it was petrified, but what you could do is ask Fickel to say whether the poop comes from a carnivore or herbivore. You know, 'cause he could do that by checking for things like seeds or bone fragments."

Randy's mouth opened slightly while he gave the human in red an incredulous stare. "Thank you, Professor Bellamy. Once again, your razor-sharp mind missed the joke completely."

"Whatever, man," answered the chastised boy. "If that was a joke, it sucked."

Randy turned to Borias and said, "I thought it was good. Did you think it was funny?"

"Pretty funny," answered Borias.

"You know what, Rankin? I just figured out your Halloween costume. You can go out as Mr. Peabody, you know, that genius dog on the Bullwinkle show. We can put some specs on your nose and staple a puppy-dog tail to your butt, and you can run around all night telling people about the pyramids. Hell, you won't even have to act."

Beyond the verbal fray, PUUK was struggling to grasp the identity of the human in red. Was his name Professor Bellamy, Rankin, or Mr. Peabody?

"Ha ha, very funny."

"Rankin, you really need to learn how to lighten up. That's your problem. If you were a chill daddy, you probably wouldn't have that scab on your face—although it looks pretty good today. Did that thing hurt?"

"It looked worse than it was. Yeah, it's pretty much gone now."

"Tell you what, hang around Borias and me for a while and learn from the masters. Whaddya say Greg-o?"

"Yeah, whatever," answered Borias.

"Nice enthusiasm there, buddy," said Randy sarcastically. "No, really, we should catch a movie tonight. Monty Python's got a new one out. Have you ever seen those guys? They're a hoot."

Rankin shrugged in agreement and said, "Fine."

"Hey, we need to get a move on finding some rocks," Borias warned. "It's starting to get late."

"Did you set a time with your mom to pick us up at the truck stop?" Randy asked.

"No, I just said I'd call her when we got there. They have a pay phone, don't they?"

"I dunno."

"How far is the walk there?" Rankin intervened.

"A mile, maybe a little more," answered Randy.

PUUK watched the three boys spread out and walk away from the rock outcropping. For the next ten minutes, the environment was void of conversation with each boy engaged in his own search and apparently caught up in his own thoughts. Occasionally, a land transporter—what was the human term—rumbled by in the distance. When he heard the first vehicle pass, PUUK reacted with the same alarm he displayed upon hearing the boys' voices. However, the roadway was much farther than the boys had been, and PUUK quickly became accustomed to the sounds of revving engines and tires rolling on the asphalt.

Because the alien was observing the boys so intently, he didn't notice one of the land transporters pulling over to the side of the road and stopping. The boys hadn't noticed either. A boisterous voice from the window of the vehicle broke their concentration.

"Hey, Winters!"

PUUK gazed across the open land and spotted a human leaning his head out the window with his hands cupped around his mouth. He couldn't discern much about the subject from this distance other than to conclude he was male. The human named Randy responded to the call by trotting over to the land transporter. PUUK suddenly remembered.

A car. Humans call it a car.

The other two humans—PUUK was now relatively certain their identifiers were Rankin and Borias—strolled toward the commotion at a leisurely pace. As he drew close to the car, Randy turned and yelled.

"It's Jerry and Totem."

The alien watched Randy reach the car and engage in some sort of discussion. Borias continued walking toward the car, but Rankin stopped. After a moment, Randy turned back to his companions and hollered, "Jimbo's here, too."

Except for those instances when Randy yelled, PUUK couldn't hear the conversation at the car. He simply observed the interaction from his hiding spot. He determined three humans were sitting in the car. Randy had identified them as Jerry, Totem, and Jimbo. They appeared to be acquaintances, but PUUK couldn't draw any other conclusions.

Eventually, Borias made it to the car, and Randy walked around to the other side. Rankin continued standing motionless in the open field. At one point, Randy raised his hand and then pointed to the west. Seconds later, he screamed out, "Ah! You bitch!" while wiping something from his face.

The humans talked quietly for a while, and then Randy left the others and rejoined Rankin. Despite being closer to PUUK, their conversation was still too soft in its tone. PUUK only caught a few of their words with "Saddle up," being the clearest. Randy gently grabbed Rankin around the head, and the two walked to the car. Once there, they entered the vehicle along with Borias. The car pulled back onto the road and drove away in the direction of Casper.

In the quiet of his tree-lined surroundings, PUUK contemplated what he'd just seen. He glanced outward in the direction of the mountain and noticed the cows grazing. He wondered if any had peed on a flat rock in the interim. He didn't know, and, honestly, he no longer cared.

What he now cared about were the humans he had just observed. Something about them stimulated his curiosity. They were—for lack of a better word—alive, so very much alive. He hadn't realized how much he missed the vibrancy of active existence. Not only were these teenage humans alive, they appeared to be somewhat crazy with lots of potential irrationality to study.

PUUK's own worldly sensations were returning, too. The unusually rich oxygen content in the Earth's atmosphere instilled him with a slaphappy giddiness and started to affect his judgment. As PUUK prepared for his return journey back to his orbiting vessel, he decided on a whim to track the land transporter carrying the six humans. Yes, they would become his study group. He would monitor the lives of the boys on whom he'd just eavesdropped. He would also study their acquaintances in the car.

What were their names? He asked himself. *Ah, yes, Totem and Jerry and Jimbo.*

The first step in the process would be to individually research each of them. PUUK would shift forward and backward in time to understand their stories. While he was at it, he would review his story, too.

BORIAS

Four ladies assembled in Lorraine Eller's home for their monthly bridge game always held on the second Wednesday. After cards were dealt and hands assessed, the bidding began.

"One club," said Helen Winters. The others followed in turn.

"One heart," called Lorraine.

"One spade," answered Clair Borias.

"Pass," said Anna Young.

"Two clubs," bid Helen.

"Pass," said Lorraine.

"Three clubs," responded Clair.

"Pass," said Anna.

"Pass," said Helen.

"Pass," said Lorraine.

Lorraine was seated facing the picture window providing an expansive view of the Eller's front yard and Meadowlark Street. Clair was seated to Lorraine's left, Helen to her right, and Anna directly across the table. Anna was Lorraine's neighbor and lived two houses to the west. Clair and Lorraine shared a dear friendship and often paired together to lead community outreach efforts for the First Presbyterian Church. At Clair's suggestion, Helen was substituting for Sheila Dixon, who was visiting her sister in Indiana.

Clair and Helen won the auction with the three clubs bid. Because Helen initiated the bidding in clubs, she became the declarer.

"Here's what I have, Helen," Clair said as she assumed the role of dummy and laid out her cards for Helen to play. "Make 'em work." Clair leaned back in her chair and watched with interest as Helen began play. Helen's non-response to her three-clubs bid seemed weird.

Clair wasn't the only one carefully observing the bridge game. Beyond Earth's atmosphere, PUUK gave the game his full attention. His surveillance consisted of three types of data gathering. Audio capture enabled PUUK to listen in on all that was being said. Video surveillance allowed him to watch the subjects in search of any non-verbal communication. Finally, telepathic monitoring intercepted the thoughts emanating from the humans.

Audio and video surveillance qualified as standard fare for alien study, but thought capture was another matter. Simple technology necessary to receive sounds and visual images was readily available. Even the humans were beginning to master these techniques. On the other hand, honing in on a person's silent mental output was tricky business. Oh, it was doable but wasn't something to be done on a whim. The primary obstacle to thought capture was the power required to sustain the process. In layman's terms, thought capture hogged resources. Being a small stealth craft, PUUK's ship was limited in the amount of power it could generate, so if he was committing to this effort he had to feel confident in his reasoning. PUUK was extremely confident.

The alien was also deeply relieved. His hasty decision to select the first humans encountered as his study group was highly irrational and if discovered by his peers could be personally damaging. The fact he transmitted his selections back to his home planet made his actions irreversible. PUUK had fretted about the consequences of his behavior since the effects of the high-grade oxygen wore off. He now had to move forward with his chosen candidates and make the best of it. Fortunately, the inclusion of HUMAN SUBJECT: GREGORY BORIAS was blossoming into a scientific bonanza.

To begin his study, PUUK briefly researched each boy's history and a little bit of his future. Then the space traveler plotted a schedule for abducting each person in the study group. PUUK felt abduction was too harsh a term. He preferred to think of his actions as borrowing some of the boys' time for scientific enhancement.

HUMAN SUBJECT: GREGORY BORIAS was the first on his list. The insight gleaned from his first examination of this unstable kid was staggering. Inconsistencies in the boy's behavior correlated beautifully to the conditions PUUK had been directed to research. The boy reeked of irrationality; his unpredictable psyche could be proudly showcased in any cerebral exhibit this side of Andromeda.

The space traveler flipped through the GREGORY BORIAS file again. The human's physical attributes all fell within the norms. He had no impairments. Intellectually, he scored exceptionally high for a human. That trait might require additional study. However, the essence of his being generated a cloud of mystery. His behavior fluctuated wildly, and frankly, PUUK was a little afraid of him.

PUUK skimmed through reference databanks for enlightenment. Sporadic cases of this condition had been documented throughout the cosmos. The greatest experts in the field had studied the specimens, hypothesized the science, and coined descriptive terms. Yet, none of these descriptions more aptly depicted the teenager's mindset than the question raised by Greg's own classmates on Earth. *Why does Borias always have such a huge bug up his ass?* Those who knew him best could tell you he was always pissed off about something.

To gain insight about the BORIAS SYNDROME, PUUK liked the sound of the term, the space traveler set out to study the subject's physical and social surroundings, hence, the surveillance effort now happening at the Eller residence. PUUK programmed incoming data from the assorted collection devices to channel onto a display panel where he could simultaneously observe the various types of information. Video of the room would be overlaid with vocal and thought output from the subjects on a real-time basis.

In his short time revolving around Earth, PUUK had made two important discoveries. First, he encountered HUMAN SUBJECT: GREGORY BORIAS. Second, PUUK observed a somewhat unique human behavior. Specifically, he noticed what humans said aloud and what they thought silently were often entirely different. This discovery justified his need to commit the ship's limited power to thought capture. PUUK sat back and absorbed the incoming data as it streamed across the display.

CLAIR VOCAL: "My Gregory is something else."

CLAIR THOUGHT: "My Gregory is something else."

ANNA VOCAL: "Oh?"

ANNA THOUGHT: "You can say that again, sister."

CLAIR VOCAL: "Have I mentioned what he's been doing?"

CLAIR THOUGHT: "Have I mentioned what he's been doing?"

LORRAINE VOCAL: "No, what's he been up to?"

LORRAINE THOUGHT: "I don't really care. I'm just happy we've made it past the phase when he felt the need to pee in my garden every summer."

CLAIR VOCAL: "He's been reading about Napoleon. He practically knows everything about the man. He's so clever."

CLAIR THOUGHT: "Unlike you, Helen. You had the Ace, King, and Jack of clubs in your hand, and you don't raise my three-clubs bid. What the hell were you thinking? We could have made game and maybe a slam."

LORRAINE VOCAL: "My deal."

LORRAINE THOUGHT: "My deal."

HELEN VOCAL: "How did he get interested in Napoleon?"

HELEN THOUGHT: "Geez, Clair, lighten up on the disapproving stare. I don't play bridge five times a week like you do."

CLAIR VOCAL: "I'm not sure. I'll have to ask him."

CLAIR THOUGHT: "But I know the inspiration didn't come from associating with your boy, Randy."

ANNA VOCAL: "Isn't Napoleon the guy who always poses with his hand tucked in his shirt?"

ANNA THOUGHT: "And wasn't he an arrogant little bastard as well? I can see why Greg likes him."

PUUK continued to carefully monitor the ladies' game as it progressed in a similar manner for the next two hours. He learned assorted information. Greg Borias and Randy Winters were friends and attended the same school. Lorraine Eller had two sons, one in high school and the other in college at Laramie. Fried food was causing Stanley Young to expel more gas than usual. Pearl Morrisey found a lump on her breast.

At 4:45 and just as the game had finished, the doorbell rang. Clair took the score pad and began summing the numbers while Lorraine walked to the door. It was Paul Winters, Helen's husband.

LORRAINE VOCAL: "Why hello, Paul. I wasn't expecting to see you."

LORRAINE THOUGHT: "Why hello, Paul. Did you just get fired or something?"

PAUL VOCAL: "Hi, Lorraine. Hello, ladies."

PAUL THOUGHT: "Tread carefully."

HELEN VOCAL: "Oh, it completely slipped my mind. Paul took off work early so we could do some used car shopping for Brian. He's a sophomore now."

HELEN THOUGHT: "Free at last, free at last."

PAUL VOCAL: "How'd the big game go?"

PAUL THOUGHT: "How much bitching am I going to have to hear in the car?"

CLAIR VOCAL: "Well, it looks like Anna had high score, and I came in second."

CLAIR THOUGHT: "I would have finished first if a certain ninny knew how to bid properly. That's the last time I invite Helen to substitute."

HELEN VOCAL: "Clair's been giving us a complete rundown of Greg's summer."

HELEN THOUGHT: "I need a drink."

PAUL VOCAL: "Fine young man."

PAUL THOUGHT: "Oh, boy."

CLAIR VOCAL: "My Gregory's a miracle."

CLAIR THOGHT: "My Gregory's a miracle."

VISUAL IMAGE: Clair is radiantly beaming while Paul and the other ladies are feigning smiles.

CLAIR VOCAL: "But I worry about him. Sometimes he gets so angry and yells."

CLAIR THOUGHT: "I do all I can to protect him from the horrible things in this awful world."

LORRAINE VOCAL: "It's hard raising kids."

LORRAINE THOUGHT: "It's hard raising kids."

CLAIR VOCAL: "Gregory is a complicated boy."

CLAIR THOUGHT: "Gregory is a complicated boy."

PAUL VOCAL: "He'll be fine."

PAUL THOUGHT: "He's a butthole."

Although he didn't appreciate it at the time, PUUK was witnessing, firsthand, a common aspect of human life. When two seemingly contradictory views of a given subject are presented, the truth usually falls somewhere in the middle. Greg Borias was a complicated butthole.

He was also a miracle at least in the eyes of Clair and her husband, Ray. After sixteen years of marriage, the Borias couple was still childless and chose not to pursue adoption. They were resigned to the situation. Then at age thirty-eight, Clair became pregnant. Nine months later, she was holding a healthy baby boy in her arms. In an instant, they became a three-person family and began their tenure in child rearing.

Besides a short bout with colic, Greg was a normal baby. He slept on his back, loved Gerber plums but spit out his peas, and took his first steps at thirteen months. He received attentive care from his grateful parents and grew up in an emotionally stable and financially secure household. As Greg progressed through his childhood, he lived a normal and relatively uneventful life. Still, he was a different kind of boy. Maybe it was due to excessive doting on the part of Clair and Ray. Maybe it was because he was an only child and spent most of his free time around grownups, or maybe it was his chemical makeup. Whatever the case, he was short on patience and long on hostility.

Greg's interest in Napoleon began two years ago in 1973. During that summer, the Watergate story ruled the media. It supplanted the Vietnam War as the headline event in newspapers and on TV.

In a nutshell, President Richard Nixon, a Republican, was running for reelection in 1972. The president's close aides ran a spy campaign against his adversaries, and in June of 1972 five men were arrested for breaking into the offices of the Democratic National Committee located in the Watergate building in Washington D.C. Over the next two years, those burglars would be linked to key members of the president's White House staff who ordered the break in. Further investigation revealed when President Nixon became aware of the Watergate incident, he participated in attempts to cover up the event. As the investigation continued, key players in the break in and cover up fell like dominos and were indicted. Finally on August 8, 1974, Richard Nixon became the first president in U.S. history to resign from office.

The details involved and even the magnitude of the situation were lost on a twelve-year-old boy. To Greg, all this attention was a constant series of blah, blah, blahs and yak, yak, yaks. In the summer of 1973, Greg and the rest of America were being deprived the privilege of viewing their daily television game shows like *The Price is Right* because of live broadcasts from the Senate Watergate hearings.

Out of boredom one day, Greg accompanied Mrs. Borias on her visit to the public library. He walked immediately to the military history section. A few years earlier, Greg had watched the movie, *Patton*. He thoroughly enjoyed the movie, especially Patton's opening monologue with the line about grabbing the Germans by the nose and kicking 'em in the ass. Like most young boys, he loved the action scenes and watching tanks getting blown to smithereens. Yet, there was something about *Patton* that made it different from the standard WWII flick. Yes, there was action, but it also contained strategy and in particular the benefits of knowing history and using it to your advantage. Greg deposited this realization deep into his memory.

As Greg leafed through a book about the United States' role in the Pacific, a male librarian placing returned books back on the shelf watched him.

"Are you looking for anything in particular?" the librarian asked courteously.

Greg hadn't noticed the librarian and jumped a little. "Uh, no. I'm okay."

"War is a fascinating subject," the librarian said. He liked to encourage young readers. "Do you have a special interest?"

Greg had now recovered from the initial shock and accepted the librarian's question thoughtfully. Having done so all his life, he felt comfortable conversing with adults.

"Generals," Greg answered after a pause. "I really like General Patton."

"He was an incredible commander," the librarian replied. "Harsh but effective."

The librarian restrained himself from pointing out a technicality in the boy's search for generals. The book Greg was examining dealt with U.S. naval campaigns in the Pacific. In the navy, the highest-ranking officers are called admirals not generals.

"Are you more interested in modern-day generals or historic figures?" the librarian asked.

"Well, I know about George Washington, Custer, and Robert E. Lee," Greg answered. "So I think I know most of the historic stuff."

"What about famous European generals? Have you studied Napoleon Bonaparte?" the librarian prompted.

"I've heard of him. Was he any good?"

The librarian chuckled. "Any good? At one time he controlled most of the European Continent. He was a master tactician in the battlefield."

Greg's interest was piqued. "Do you have any books about him?"

"Several. Let's take a look." The librarian led the boy to two shelves of European history and recommended a couple of selections.

Greg picked one and immediately became engrossed in the military escapades of Napoleon. Wow! This guy was an ass kicker. He chased the Austrians, Prussians, and Russians all over the place. When Greg learned the board game, *Stratego,* was based loosely on the Napoleonic period, it immediately became his favorite. While the battle descriptions initially captured his attention and imagination, Greg began to appreciate the exceptional skills of Napoleon over time as a civic thinker and a great motivator of men. Here was this outsider from Corsica, who came in and became the Emperor of France—outsider. Greg often felt alone like an outsider, too.

When Greg's mind wasn't occupied with the study of Napoleon, it was often directed toward stirring things up at school. He liked getting under people's skin, especially teachers. Too often, he found himself bored by the pace of a lesson or annoyed by the sheer stupidity of his classmates. He felt the need to spice things up a bit. In this setting, he wasn't Greg or Gregory. He was Borias.

As the first day of school drew near in August, Borias pondered the mischievous possibilities. He wanted to make a splash, but it had to be something unique, something clever.

This was ninth grade, damn it! He had put in his two years of bullshit. It was time to rule the school.

On a bright Monday morning, excited chatter filled the halls of Collins Junior High when the 8:20 bell rang. Students traipsed to their lockers before filing into their designated homerooms. Borias stopped at his locker to drop off his toothbrush and toothpaste. He had been wearing braces for the past eighteen months, and like most kids fitted with the gear he hated the damn things. However, he was fastidious in his oral care almost to the point of it being an obsession. After shutting his locker, Borias entered science class, grabbed a seat in the back of the room, and waited to hatch his plan.

Barney Fickel, a twenty-year vet in the teaching ranks, with the last six being earth science, walked into the room just as the 8:25 bell rang.

"Ahem!" Mr. Fickel croaked and pointed to the loudspeaker in the upper corner of the classroom. On cue, Principal Andersen began the morning announcements.

"Welcome to Collins Junior High. We hope you enjoyed your summer and are now ready to challenge yourself this school year. Morning announcements will be made daily at 8:25 during homeroom, and first hour classes begin at 8:30. Today's lunch will be chicken fried steak, mashed potatoes with gravy, green beans, iced chocolate cake, and milk."

"Far out!" yelled a voice in the class. Chicken fried steak was popular with most of the students. Mr. Fickel looked in the direction of the voice and shook his head disapprovingly.

Principal Andersen continued, "Individual lunches are sixty-five cents, or you may purchase a lunch ticket with five punches for three dollars. For anyone who didn't get a picture taken during schedule pick up, the photographer will be here tomorrow after school. While you were on break, the gymnasium floor was refinished. Please remember to only wear tennis shoes when you are in the gym. Finally, I'd like our returning eighth and ninth graders to help me welcome the incoming seventh-grade class and all the other students new to Collins. Let's make it a great year."

Borias noticed a guy sitting in the front who responded to the principal's welcoming of new students. Borias hadn't seen the kid before and disliked him immediately.

Mr. Fickel stood front and center in the class. "Okay, class, my name is Mr. Fickel, and this is first period, ninth-grade earth science along with homeroom. Does everyone belong here?" Mr. Fickel glanced about the room and continued, "If you don't, please get up and go to where you do belong. We won't laugh at you—very much."

Most of the class chortled, but Borias leaned back and silently assessed the instructor.

Barney Fickel, Borias thought to himself. *Nice name! I don't think you belong here. I think you should have your butt parked on a Main Street bench in Mayberry eating goober peas.*

Mr. Fickel took two steps back to the front of his desk and twisted his body to pick up a notebook. Then he leaned against the desk almost achieving a sitting position.

"My policy is to let you sit where you want unless talking or horseplay becomes a problem. So, I'll be passing around a seating chart for you to fill in, but today I'll take roll out loud. Say 'here' when I call your name."

He began, "Kelly Adams."

"Here."

"Debbie Atkins."

"Here."

"Rankin Bellamy."

"Here."

"Greg Borias." Mr. Fickel pronounced the last name as Bo-RI-us. There was no response. "Greg Borias?"

"It's BOR-ee-as. My name's pronounced BOR-ee-as," said Borias correcting the teacher.

This was news to Rankin, and he turned to look at Randy, who was also in the class, for clarification. Randy's shrug of the shoulders indicated he was also unaware of the name differentiation. During his life, Borias could never claim to have had a best buddy, but Rankin and Randy were friends of his and knew him about as well as anyone.

"BOR-ee-as?" Mr. Fickel repeated seeking confirmation.

"Yeah."

"It wasn't BOR-ee-as last year," Mr. Fickel pressed. Fickel knew this statement to be a fact as Borias had been in his study hall group.

Borias never flinched. "It is now." Then he added, "Sir," as an afterthought.

Mr. Fickel continued to stare.

Borias met the stare and after a moment offered an explanation. "My family's had some trouble, and we've decided a different pronunciation of our name would help give us a fresh start. Kind of like the Theismann-Heisman thing." Borias was referring to an incident a few years back in which Notre Dame quarterback, Joe Theismann, at the time pronounced *Theesman*, changed the pronunciation of his name to *Thighsman*. The rumor floating around this name change said it was part of the quarterback's public relations campaign to win the Heisman trophy, an annual award given to college football's best player.

Fickel wasn't buying it but decided to let it go. "BOR-ee-as. Okay," the teacher said with a wry smile.

Neither Rankin nor Randy shared second period with Borias, so they weren't there to witness the altercation arising between Borias and Mrs. Tuttle, the English teacher. Borias again claimed a mispronunciation of his name, but this time he offered Ba-REE-us as the true pronunciation. Mrs. Tuttle said that was nonsense. She knew Greg's mother and had never heard her use that pronunciation. Borias told her to take it up with his great grandfather, whom he met for the first time during the past summer at a family reunion. He had just come over from the old country, and he's the one who insisted on Ba-REE-us.

Borias and Randy reunited again in third-period Civics, but Randy wasn't focused on Borias. Instead, he was annoyed with the limited choice of seats when he entered the class. He knew his friend, Jerry, would be in this class, but all the seats around Jerry had been taken. Randy settled for a spot across the room.

Borias sized up his next adversary. Mary Shomborg was not only new to Collins Junior High; she was new to the teaching profession, a complete rookie. During roll call, Ms. Shomborg barely uttered a *B* sound before Borias delivered a preemptive strike.

"It's Bo-RI-us!" he screamed.

Ms. Shomborg was taken aback. Trying to make it through the simplest of rituals without incident on her first day, she apologized. "Bo-RI-us. Thank you, Greg. Do you prefer to be called Greg?"

"Yes, I do. Thank you kindly," mocked Borias. He was performing at his snotty best.

Ms. Shomborg continued taking attendance. About three quarters of the way through, she called, "Girard Prentice."

Jerry's jaw dropped as his face turned white. "It's Jerry," he quickly corrected her.

"Oh, yes. There's a note here. It says to call you Jerry," Ms. Shomborg replied with an innocent smile. "I'll be sure to remember."

Jerry fumed. *Yes, there's a note there*, Jerry thought to himself. *You know why there's a note there? It's because I dragged my mom to school before I started seventh grade to make sure every teacher in the building knew how much I hated the name Girard and that they were never to use it. No one was supposed to know about Girard, and now thanks to you everyone knows about Girard.*

While Ms. Shomborg finished with attendance, Jerry calmed himself. He began to hope, *Maybe no one else picked up on it. Just be cool, and it will pass. And don't look over at Randy. Whatever you do, keep eyes away from Randy.*

Thirty seconds later he glanced at Randy. With his head tilted and eyebrows raised, Randy's *Really?* expression told Jerry this incident didn't pass unnoticed.

"Damn it," Jerry muttered under his breath.

The remainder of class crawled along uneventfully as Ms.

Shomborg asked only a few questions and made a point to stay away from Borias. When the bell rang, Borias dropped off his books, picked up his dental gear, and headed to the cafetorium—the large, open room with a stage that served as both the cafeteria and auditorium—for lunch. He had executed his pronunciation scheme with expert precision, and Borias believed he maintained the upper hand in all his classes. Now he wanted to hang out and collect some *atta boys* from the guys.

By the time he exited the chow line with his tray of food, most of the guys were already sitting at the table. *All right men,* he thought to himself as he took a seat, *let the praise begin. Was I a genius or was I a genius?*

No one mentioned the name gag. Instead, everyone argued about the best lunch at school. One camp held firm to chicken fried steak while the other side advocated chili with the cinnamon roll. Finally, Borias made eye contact with Randy.

"What's up with the name?" Randy asked.

"Identity crisis," Borias answered smiling. *Now we're getting somewhere,* he thought.

Randy accepted Borias' answer by nodding his head and then returned his attention to the mashed potatoes on his plate. Borias waited for the others to acknowledge his efforts, but there was nothing. Borias watched Randy as he spotted Jerry sitting across the table and down a few spots. Randy cleared his throat and in a booming voice directed his next comment toward Jerry.

"Excuse me, Girard," Randy began, "but could you please be such a good fellow as to pass the salt and pepper." Randy knew he had Jerry by the balls.

Jerry grabbed the salt-and-pepper shakers. With a killer's stare focused on Randy, he handed the shakers to Jimbo DuPree, who in turn passed them to Randy. Randy was completely unfazed by the, *you're-dead* look. Instead, he clutched his heart and summoned his best Italian mama impersonation.

"Girard, why you hurt me like this with these secrets you keep?"

Jerry was toast, and he knew it. Still, he tried to downplay the situation.

"Girard, Jerry, it's the same thing," he responded.

Randy quickly corrected him. "Oh, no, they're not even the same thing."

"Not even," echoed Rankin with a huge smile. He had been watching silently and was grateful not to be the person, for once, who was catching grief. He popped some Spree candies into his mouth. Then most of the guys in the cluster chimed in with various comments about Jerry's formal name.

Borias watched this scene unfold and couldn't believe what was happening. Was Girard really usurping his time in the limelight? Of all the stupid things in this world, this turn of events ranked right up there. Girard? That's not even funny let alone clever.

Borias' disposition immediately soured. When the bell rang, he stormed through the halls and back to his locker. He was so upset he didn't bother with his name charade during fifth or sixth period. His mood grew nastier as the minutes passed.

The final class for Borias was seventh-period Latin. When the day began, he was actually looking forward to this course, but now all he wanted to do was sulk in the back of the room.

Mr. Hatcher greeted the class with his *carpe diem* speech and then proceeded to spend over half the period lecturing about the Roman's influence over the Western World. Toward the end of class, Mr. Hatcher handed out blank maps and time-lines that were to be completed as a first assignment. He doled out the exact number of copies for each row of students and instructed them to take one of each and then pass the rest to the next person.

Christine Larson sat in front of Borias and used this oppor-tunity as a chance to greet the boy.

"Hi, Greg! How's it going?" She should have stopped at "Hi, Greg."

Being asked how it was going when he was hacked off irri-tated Borias even more. However, a simple *shut up* wouldn't do. Borias had a knack for making his comebacks memorable. He took the papers Christine offered and then paused for a moment while he studied her.

"Christine, I never noticed before but you have the most amazing *ottabeon* hair," Borias said in earnest.

Christine bit. She didn't comprehend the adjective Borias used, but the tone of his voice was complimentary. Compliments were always welcome.

"Thank you, Greg. Ottabeon?" she asked.

"Yeah, otta-be-on a dog's ass," Borias snarled as he deliv-ered the punch line. "Please don't tell me you spent all summer working on *that* look."

The stunned girl reeled back in her seat and began to process what had just been said. She hadn't spent all summer working on *that* look, but like most girls she had spent an inordinate amount of time fussing about her appearance especially for the first day of school.

The insult fully soaked in after a few moments, and then Christine began to wilt. Crocodile tears streamed from her face. She hid her eyes and waited for the bell and the longest eight minutes of her life to pass.

A few of the other students heard Borias and Christine's exchange. Lance Chitwood snorted in laughter, and Cindy Mathers frowned in utter disgust.

"You're such a bastard," she said.

Jimbo shook his head in disbelieve and said, "Man, that's cold."

A few minutes after the Borias burn, Mr. Hatcher noticed Christine in distress. He spent the remainder of class trying to ignore her.

"Not on the first day," he thought. Mr. Hatcher was retiring at the end of the year, and one of the things he wouldn't miss was dealing with emotional teenage girls.

When the bell rang, Cindy hurriedly escorted Christine out of the room. Borias sat in his chair for a few moments, finally gathered his items, and stood. He thought dumping on someone might help him feel better. It didn't. He briefly stopped at his locker and then exited the building through the front doors. From school, he followed the same path home he had taken during the past two years, and he took it in the same manner—alone.

TOTEM

With the breeze mostly at his back, Henry Sweetgrass climbed out of the narrow draw and up the embankment. The brown earth, no longer muddy but still soft with moisture from a recent rainstorm, cushioned each step as his feet dug into the hillside on his ascent. Still, Henry's big toe throbbed, and he winced in pain each time his right foot pressed into the ground. Unconsciously, Henry thanked Mother Earth for her spongy surface which eased the difficulty of his journey. Unconsciously, too, he gave thanks for the pulsating pain which took his mind off other thoughts.

Many moons ago, an opening his father often used to begin a story, Henry and his family moved to Casper from the reservation. He was a young boy, eight years old to be exact, but he wasn't a small child. Henry stood taller than most of his classmates, a condition fortunate for Henry because it deterred any teasing that might have come his way. Henry was an Indian boy in a mostly white school, but second graders by and large were an accepting group. It wasn't his race that had potential to draw taunts but rather his performance in the classroom or, more specifically, lack thereof. Henry wasn't stupid, but he was overwhelmed in this formal-learning climate, and the fact he arrived to his new school in February didn't help. He departed a culture in which children simply listened to stories and replaced it with one in which people read books. His old friends on the reservation were still learning their A, B, C's, but here his classmates all carried readers with stories about Dick and Jane or the adventures of a mischievous little dog named Tip.

Despite his struggles, Henry passed into the third-grade class of Mrs. Irene Bateman. Unfortunately, his reading issues persisted and even deteriorated. After three lengthy conferences with Mr. and Mrs. Sweetgrass, everyone agreed Henry should remain in third grade another year.

The normal practice for a student repeating a grade was to assign him to another teacher working that same grade. Trooper Elementary presented two alternatives to Mrs. Bateman; however, Mrs. Bateman wanted more time to work with Henry because she was intrigued with the challenge. She was confident she understood the boy and his learning needs. The seasoned teacher's determination to work again with Henry trumped normal protocol. Further, Mrs. Bateman served as the senior member of the third-grade staff, and, well, seniority has its privileges.

Had he bothered to account for key periods in his life, Henry would have listed the extended time with Mrs. Bateman. She liked him, and he liked her as much as he was going to like any teacher, and, yes, she did help Henry overcome his reading deficiencies. Henry would never be a great reader, but at least his skills were now adequate and enabled him to advance through the grades. Also, given all the people in Henry's world, it was Irene Bateman who pinpointed and verbalized the defining characteristic of Henry Sweetgrass.

"Your son spends more time inside his head than any child I've ever known," she told Mrs. Sweetgrass during the spring parent-teacher conference.

Henry now only vaguely remembered his second go round in third grade. He recalled being happy because Jerry and he were in the same class, a selling point his folks used when breaking the news about having to remain in third grade. He also remembered the long rectangular window and the peaceful outside world it revealed. That world captured his attention for minutes at a time, broken only by Mrs. Bateman's repeated requests for him to return to his assignment.

Except when his desk was turned to face the inward wall of the classroom—a tactic Mrs. Bateman employed on multiple occasions to retain the boy's attention—Henry was afforded a nicely framed snapshot of the outdoors. Centered in the window's frame stood an elm tree whose mere presence offered Henry great comfort. Seasonally, the framed tree dressed itself differently with varying colors of foliage or no leaves at all during the dead of winter. In December and January, the sun stayed low enough in the sky to partner with the tree in the confines of the window for a short time each day. Occasionally, a sparrow or possibly a robin might land on a branch enhancing Henry's scenic scape. The finer details might vary, but always within the window's frame remained the elm tree. Whenever he looked out, it was there, a known constant. He could count on it.

To his classmates, Henry became known as the quiet giant who stared out the window. He was already tall for his age, but now with an extra year on the rest of the grade, he towered over them. Ironically, the tiniest person in class changed the identity of Henry Sweetgrass.

Carrie White was as teeny as Henry was large. She seldom spoke but on the few occasions when her mouth opened, she didn't talk; she squeaked. Because Carrie was heard so infrequently, when she did squeak everyone noticed. And so it happened during a show and tell period as Tracy Barber shared her family's picture slides from their past summer's trip. Midway through the presentation, an unusual photo appeared. In it, Tracy and the rest of her family were posing next to this enormous statue. The statue looked like a giant log standing on its end with a bunch of different faces carved out, one on top of another.

"How interesting!" exclaimed Mrs. Bateman. "Tracy, what is it?"

Brimming with knowledge, Tracy replied, "This is a totem pole. It is a very important part of Indian culture."

Brady Johnson turned to Henry. "What do you do with it?"

Henry was speechless at first and then blushed in embarrassment. "I don't know," he stammered and then added, "I've never seen one before."

Mrs. Bateman jumped in. "Totem poles are found in the Pacific Northwest and are made from giant trees found in that area. Brady, go over to the map and show us the state of Washington and the area above it."

Brady stood and walked to the map. He first pointed to Texas and then Pennsylvania. Brady was never going to win a geography competition, but he finally found the requested spot with some assistance from the teacher.

Mrs. Bateman continued, "There are many different types of Indians in North America, but each tribe uses the natural materials found in its region. Henry belongs to a plains Indian tribe, Sioux, I believe. Is that right, Henry?"

Henry nodded.

"Wyoming is part of the plains, and tall forests don't grow around here, so plains Indians don't have totem poles," the teacher explained.

"So, there are different types of Indians?" Danny Plimpton asked.

"Yes, just as there are different types of white people. Your families probably came from different countries such as England, France, Germany and other places."

Many of the students had never considered differences within a race before, not the least of which was Henry. He cautiously smiled at Mrs. Bateman, grateful for her rescue. She returned the smile. Then out of the blue, Carrie White squeaked.

"So, if we painted some faces on Henry's tummy, he could be a totem pole," she proposed.

The class burst out in laughter, and once more Henry felt his face go flush. Denise Carter thought it was so funny she shared it with her family including her brother, Monte, at dinner. Monte had shared Henry's third grade class the previous year and knew him well enough to hang out on the playground. At recess the next day, the boys were choosing teams for football. As one of the team captains, Monte selected Henry with his second pick.

"I'll take Totem Pole," Monte announced.

The guys in Henry's current class understood the reference and laughed. The guys who weren't in Henry's class had no idea what was going on but laughed anyway. Henry winced with discomfort but walked over and stood next to Monte. In the weeks and months to follow, Henry's friends condensed Totem Pole to Totem bestowing the boy with a new nickname.

Back in the present, Henry crested the hill and caught his breath. Summer was nearing its end, and the wind blew in bursts against the tall, straw-colored grass. The stalks of grass yielded downward with each breath of air and then snapped back during a pause. Stubbornly, the cactus and yucca plants growing in the vicinity refused to heed to the wind's whims. The boy inhaled the fresh air and relaxed. He had reached his comfort zone. Closing his eyes, Henry remembered happy times in the past.

His mind wandered back to his family's first house in Casper. Located on Wrangler Street, just off the old Yellowstone Highway, the ranch-style home was simply designed, but the sights, sounds, and smells from that house were embedded in Henry's memory. He heard the distinct chatter of his mother's pressure cooker as it readied an elk roast for supper. He recalled Sunday evenings in the living room as the family watched *Wild Kingdom* and then *Bonanza*, but his favorite memory involved the Sweetgrass boys and their bedtime ritual.

Like most children, Henry and his brother, Edward, gave in to their imagination whenever the lights went out. They feared monsters, spiders, and all the other creepy things lurking in dark corners. Mr. Sweetgrass, who was in charge of bedtime preparations, knew all too well the boys' anxieties as one or the other would frequently pay a midnight call to the bedroom he and the Mrs. shared. Weary from all these unsolicited visits, Mr. Sweetgrass devised a remedy.

"Are you ready for Happy Dreams?" he asked the boys.

They responded with hoots and giggles. Mr. Sweetgrass invented Happy Dreams as a preemptive remedy against the Bogeyman, and it became a mainstay in the nighttime tradition right after teeth were brushed and prayers had been said. Henry, Edward, and Mr. Sweetgrass piled onto one of the twin beds in the boys' bedroom. William, the eldest Sweetgrass son, occasionally poked his head into the room. Because he was much older, he never directly participated, but he couldn't help being drawn in by the infectious nature of the ceremony.

During Happy Dreams, Mr. Sweetgrass told a story consisting of only happy thoughts. The story might go as follows:

Many moons ago there was a young Sioux brave who woke to warm sunshine on his face. He walked to the nearby stream and drank cool water to relieve his thirst. After drinking, the young brave looked up and saw a spotted fawn in the brush. The fawn beckoned the young brave to follow. They walked over a hill and saw many chokecherry bushes. The young brave stopped and tasted a berry on the first bush. It was ripe and juicy. He stopped and ate another berry at next bush, but the young brave noticed this berry tasted slightly sweeter than the first, and it was less round in shape. At the third bush, the young brave enjoyed an even sweeter berry. He noticed this bush had two colors of berries. Some were red and some were purple, but both were sweet, and neither was shaped like a berry. The young brave looked down to the end of the row, and

*there he saw a bush with berries of all colors. "What is that bush?"
he asked the fawn. The fawn replied, "That is the jellybean bush,
and it has the sweetest berries of all." The young brave ran to the
jellybean bush and tasted each color. He ate and he ate until he
had a terrible bellyache.*

Mr. Sweetgrass paused, and the boys immediately protested.

"Father, these are Happy Dreams, and bellyaches are not
happy," they cried.

Having stated their case, the boys sat back and waited
for the finale, as they knew the ritual well. The face of Mr.
Sweetgrass grew serious, and he continued.

*The spotted fawn did not like to see the young brave in pain, so
he gave the boy a can of root beer. "Drink this as fast as you can,"
the fawn told the brave. The brave took the root beer and chugged
it. He felt an enormous bubble in his belly, and then he farted. He
farted and farted until his bellyache was gone.*

By this time, the boys were jumping on the bed and squealing with laughter. They hopped on Mr. Sweetgrass and he
tickled them. Tickles slowly turned into cuddles and hugs, and
the room eventually grew quiet. Mr. Sweetgrass asked his sons
if they had tried hard in school that day, and they assured him
they had. The devoted father tucked each boy into his respec-
tive bed and turned out the lights.

Henry hadn't thought about his family's togetherness for some
time. He remembered his family shopping along Main Street on
Saturdays. They generally kept to themselves but responded back
politely to any passersby greeting them. He imagined how his
family must have looked to someone else. *They looked normal and
respectable*, he thought, *just like any other family.*

Truth be told, those people in Casper acquainted with the Sweetgrass clan would say that other than looking like Indians, they were pretty normal. Although they wouldn't say it aloud, those same folks thought being normal meant acting white, not that many of Casper's citizens had enough in depth exposure to the Indian culture to make an informed comparison.

Yet, every family possesses its oddities, and the Sweetgrass' were no exception. Their peculiarity centered on the formality in the way they regarded their names. Casper was full of Moms and Dads, but Henry's parents were always Mother and Father. The formality pertained to the boys as well. An onlooker might see the three brothers walking down the street one day and say, "There goes Billy, Eddie, and Hank," but he'd be wrong. They were always William, Edward, and Henry. Mr. Sweetgrass called them his three kings. So, when Henry was renamed Totem by his friends, it caused much consternation on the Homefront.

"Your name is Henry," Father told him, but Mr. Sweetgrass only had so much control over the situation. In time, Henry became accustomed to dual identities—Henry at home and Totem at school.

William encountered a challenge to his name as well. A local sportswriter penned the name William *Straight Arrow* Sweetgrass after William swept the long jump and triple jump events at the regional track meet during his junior year of high school. In his own cutesy fashion, the sportswriter applauded himself for not only concocting a nickname that visually described William sailing through the air but also one that captured the boy's serious demeanor.

"I hate the name," William complained to his father. "It's so patronizing."

Interestingly, Mr. Sweetgrass regarded William's nickname much more lightly than he did Henry's. "Roll with it," he advised William. "Hey, you could be called a lot worse."

William heeded his father's advice and never protested when a voice in the crowd would call out, "Go get 'em, Straight Arrow." Along with his serious nature, William displayed amazing self-discipline, and these combined qualities made William appear much older than he actually was.

Henry regarded William as the perfect big brother. When Father was unavailable or in a cross mood, William filled in as a second dad. He helped Henry learn cursive and showed him how to clean a trout. A grin spread on Henry's face anytime his big brother mussed his hair, and when William let him ride on his shoulders, well, Henry doubted life could get any better. Because of their age difference, there was no rivalry. William was Henry's favorite adult, the one who made no demands but could always be counted on in a pinch.

A Marine recruiter took note of William's maturity, too, and he was always grateful when he could be truthful in his sales pitch. "You know, son," the recruiter told William, "I see leadership ability in you. You're officer material plain as day."

The recruiter didn't have to do much selling. Joining the military had been on William's mind since he turned sixteen. Father was a proud veteran of World War II, and William felt duty bound to follow in his footsteps. America was engaged in a fight over in Southeast Asia, and serving is what you did for your country, cut and dried. William appreciated the recognition for his leadership potential, but officer training took time. William wanted in the action *ASAP*. He knew he could enroll in officer training after he picked up some experience in the field.

William's attitude regarding patriotic duty and the Vietnam conflict wasn't widely shared among his peers. Casper would never be a trend-setting town, but the anti-war movement had been established throughout the United States long enough so that even conservative bastions like Wyoming were affected. Ironically, instead of being swayed by the prevailing teenage sentiment, William became more resolute in his beliefs.

Those damn long hairs are nothing but a bunch of cowards, William thought. The sportswriter had hit the nail on the head. William was, indeed, a straight arrow.

Mr. Sweetgrass felt torn about his eldest son's aspirations. He took immense pride in his son's character and courage, but he also knew the potential consequences of battle. His trepidation about William's future caused many restless nights. Mother remained silent. After much discussion between father and son, William enlisted in the Marines.

Soon after graduation in the summer of 1969, William said goodbye to his family and headed to California for basic training where his leadership abilities stood out. Racial tensions between blacks and whites were running high throughout America, and this mistrust carried over into boot camp. For once, William's Sioux heritage worked in his favor. Because he fit into neither camp, he was regarded as a neutral party. William used that neutrality to diffuse some fights in the barracks.

While race played an initial role in creating opportunities

for William to shine, his outstanding character ultimately drew the respect of his peers. He was disciplined, knew how to handle a gun, and offered assistance when someone needed a hand. In short, William had his shit together. His drill instructor recognized these qualities, too, and appointed him as a squad leader.

A few months later, William's unit shipped out to South Vietnam where William led a fire team. During the early days of his tour, William wrote to his family when possible; however, as his combat experiences accumulated he felt less inclined to communicate anything to anyone. William was getting exactly what he had asked for when he joined the Marines, and there were many times he wished he could give it back. He was a skilled, courageous, and patriotic warrior. As a soldier, he exemplified the total package, but even the best need to have luck on their side. On one sweltering day in the jungle, William did not.

Henry gingerly knelt down atop the hill and pushed himself into a sitting position. He stretched his legs outward still cognizant of the pain in his toe. He considered removing his shoe to have a look but decided against it because he wasn't sure he could put his shoe back on. He returned to his memories.

William never came home. Mother never came home, either. He thought about her. She had been gone more than a year now. He remembered the day she left. She was absent from the house when he returned home from school. Father arrived from work at dusk, and there was still no sign of her. He recalled Father's worried face entering the doorway after returning unsuccessfully from his search around Casper. Two days of tortured hell passed before they received a phone call from his grandmother. Mother had taken a bus back to the reservation in Montana. Father immediately drove north, but he came back alone.

Henry picked a piece of grass, started tugging at its end with his teeth and continued his recollection. He decided Mother had checked out long before she climbed onto that bus. When? Was it William's death? Maybe, but it seemed to be more than that. Did the move to their second house on the outskirts of town have something to do with it?

Henry was too young to notice at the time, but of all the family members Mrs. Sweetgrass was the only one who never assimilated into Casper life. She didn't know how to drive, and she only went downtown when the rest of the family could join her. Even when shopping for groceries, she was always accompanied by either her husband or sometimes by William, once he began driving. When the family first arrived in Casper, the boys were small and kept her busy. Sometimes, she would walk Edward and Henry to school, but she never mingled with the other moms. As the boys grew, they demanded less from her. At first, she appreciated some quiet time. She especially enjoyed reading the sports page over and over again whenever William had participated in a track meet; however, as time passed she became lonely. She missed her friends, extended family, and the comforting surroundings of the reservation in Montana.

The news of William's death devastated Mrs. Sweetgrass. Instead of grieving together, each family member coped with William's loss separately. Father kept to himself, and the boys' demeanor changed to a somber tone. It seemed to Henry that this, too, was about the time when Edward began causing trouble. Depression set in the mother's heart and grew worse over time.

As Henry was completing sixth grade a few years later, Mr. Sweetgrass announced they would be moving during summer break to a new house on the edge of town. He scored an unbelievable deal he told the family, and he had big plans for the new place. The new place included two acres of property. Mrs. Sweetgrass had no idea what they were going to do with the property. Besides Buck, their dog, they didn't own any animals. However, Mr. Sweetgrass assured her that he had big plans. The house at the new place was actually a slight downgrade from their current home, and that's why Mr. Sweetgrass scored such an unbelievable deal. His vision was obviously distorted because everyone else clearly saw this house as a fixer-upper. Not only did the house require some major work, it sat two miles outside of town. They were in the boonies.

The family's move only added to the isolation the distraught woman already felt. Then one day, Mother was gone. A few months later, Mrs. Sweetgrass wrote the boys a letter trying to explain why she left. She rambled but the gist of the message was that she returned to the only place where she knew comfort and security.

Surprisingly, Henry never hated Mother for what she did. He didn't understand her actions, but he accepted them. With his imagination, he created the strangest image of his mother and then filed it into his memories. The image involved a square frame and on each side stood one of the male family members: William above, Father and he to the left and right, and Edward below. Around the perimeter of the picture, the skies were clear, but in its middle snow fell hard. Mother was standing in the center of the image completely lost in the storm.

Henry leaned back on his elbows and let his long black hair drape behind him. From the hilltop, Henry viewed part of downtown Casper. This vantage point was more stirring in the evening when the lights twinkled below. In the daylight, Henry preferred to look to his right where he could see the North Platte River bend in the foreground with the west end of Casper Mountain as a backdrop. About a quarter mile below where Henry rested, a dirt road passed through the opening of a barbed wire fence. Two tall wooden posts marked the entrance, and a cattle guard had been inserted across the road. Henry tilted his head so that the river's bend lie visually centered between the two posts. Then, he simply watched.

The vista Henry framed with the posts was serene. Occasionally, a duck would land and float about in the water. In the absence of any waterfowl, Henry stared in random contemplation at the river's ripples. Putting boundaries around a natural setting was Henry's coping mechanism. It was why he had taken the long, and today anyway, painful hike to get here. He often felt if he could establish similar boundaries for his life, he'd be in much better shape.

After a while, Henry snapped back into reality. He promised Father he would do the laundry today. Hopefully, Edward would be gone by the time he returned to the house. Edward had turned into a piece of work. It seemed to Henry somewhere along life's journey, there was a huge fork in the road. William had picked one path and Edward the other. Henry used to be certain he would follow William, but now he wasn't so sure and that scared him.

With only two years separating the brothers, Edward and Henry were constantly battling. Recently, Henry had closed the size gap, but Edward possessed a crazy mean streak giving him the drive to push an argument beyond Henry's more reasonable limits. Growing up, Edward lived in William's shadow. Teachers, coaches, and neighbors dealt harsh criticisms when he didn't measure up to William's standards. Mother's departure and Father's preoccupation with his own misery effectively left Edward on his own. At age seventeen, he responded by quitting school and loitering around town.

He had been arrested twice in the past year, once for vandalism and the other for underage drinking combined with an open container. Undeterred, Edward now was moving into serious action. He had started dealing weed in small quantities, nickel and dime bags, but he was anxious to move up the ladder.

"Sweetgrass, that's my name and my product," he used to brag. Perhaps he should have tried his hand at marketing.

Edward still lived at the house, but his appearance there was sporadic. According to Henry, it was a good night when Edward didn't come home. An acceptable night was one in which Edward came home but didn't make it past the living room. He would crash on the sofa with the TV still playing long after the television stations had signed off for the night. But on those occasions when Edward was restless, well, those were bad nights. Last night was a bad night.

One of Edward's favorite games was to drip Tobasco sauce

onto the lips of his sleeping brother. Once, he put a drop of the hot sauce on Edward's eyelid. Last night, however, he took a cinder block and put it in the doorway of Henry's pitch-black bedroom. In the middle of the night, Henry went to relieve his bladder. On the way to the bathroom, he kicked the cinder block with his toe and immediately coiled in pain. He was pretty sure he had broken it.

From his console, PUUK patiently observed the boy resting on the hill. Unlike HUMAN SUBJECT: GREGORY BORIAS, this human specimen wasn't discharging bolts of irrationality or any other type of outward emotion. HUMAN SUBJECT: HENRY SWEETGRASS conducted his life in an introspective fashion. During PUUK's surveillance, the only noted oddity about the teenager was the assignment of dual names.

Beyond the names, PUUK had only succeeded in collecting a minimal set of facts about the boy. The human was nursing some type of foot injury and spent an inordinate amount of time residing in a contemplative state—oh, and one more. He had been the driver of the land transporter on the day PUUK first spotted his study group.

Still, something about this being intrigued the alien. His skin coloring was darker than his companions, and his physical movements were more controlled, less spastic. Psychologically, the boy remained a mystery, a condition PUUK planned to address during his upcoming examination.

Startled by a growing rumbling, Henry turned and watched a pickup raise dust on the dirt road as it approached from the west and ultimately pass through the gate. He didn't recognize the vehicle and wondered if its driver had seen him. If he did,

Henry wondered whether the driver might ask himself why the boy on the hill wasn't in school. It was a school day, after all.

Henry was serving his third day of a weeklong suspension. It was all so stupid! Henry couldn't believe it happened. The previous Friday, Henry played dodge ball in gym class. Gym highlighted his school day because he was athletic and tended to perform well in most sports. The game played out as an even contest and was getting down to the nitty gritty. Only Billy Martin and he remained on his team. The other side had dwindled to Jerry and Roger Caldwell, two solid players. Yup, it was Billy and Henry against Jerry and Roger—and Rankin.

He didn't notice Rankin. Who would? It was gym class, and it was Rankin. Henry's oversight was his undoing. While cocking his arm and taking aim at Roger, Rankin stepped in from Henry's blind side and literally at point blank range flipped a ball into the small of his back. Henry was stunned at first, and then to everyone's surprise he blew into a fury. He quickly stepped toward Rankin and planted his forearm into the smaller boy's chest knocking him to the ground. As Rankin tried to wriggle away, Henry pushed his head down against the floor applying a strawberry rub to the side of Rankin's face.

Without thinking, Henry began swinging wildly at his victim and might have hurt Rankin badly if the gym teacher hadn't acted quickly. Mr. Osterhagen was a thirty-something man with a thick middle and a slight German accent. When he was out of earshot, the students referred to him as Herr Potato, a name Borias coined last year. However, at this moment, Mr. Osterhagen displayed the athleticism of his youth. He pushed Henry off Rankin and then applied a full nelson until Henry submitted.

Henry was immediately suspended. He deserved to be, and he knew it. What surprised Henry was the length of his suspension. He expected two days, maybe three, but a whole week? After thinking about it, he decided the severity of his punishment was due to his victim's identity. Rankin had been bullied in a very public incident last spring. Collins Junior High was putting its foot down. It wasn't just school administrators who found his actions despicable. Practically the entire student body thought Henry was a jerk. Even his friends were riding his case on this one.

This situation reeked of irony because prior to the encounter Henry was barely aware of Rankin. He wasn't a friend; he wasn't an enemy; he wasn't even a factor in Henry's life. Henry knew who he was, and okay, if he was going to be truthful he did consider Rankin to be a bit of a wimp, but Henry wasn't out to get him. It's just that getting knocked out of the game by Rankin, of all people, happened to be the one frustration in a recent series of frustrations that made him lose it. He never used to lose it.

What's happening to me? he wondered. Henry glanced back at the path that led him to his meditative spot. He now had to retrace his steps to return home. A grim smile appeared on his face because it suddenly dawned on Henry his immediate journey reflected his life in general. He was headed downhill and against the wind.

RANDY

On the television screen, the national correspondent finished his report. "Laos now becomes the third nation in Southeast Asia to fall under communist control this year." Randy hadn't thought about Vietnam for a while, but the TV report brought back images of Saigon's fall last spring. He vividly recalled a helicopter and other large equipment being pushed off an American naval vessel and into the South China Sea to make additional space for the last waves of Vietnamese refugees. Randy could still see the throngs of people at the gates of the U.S. Embassy; they desperately sought escape from their fallen country and were fended away by Marines.

As the newscast switched back to the local station, Randy's war images quickly faded and were replaced by the news anchor's tease, "Get ready for an unusually hot weekend. Weather is next after our commercial break." This announcement struck Randy with blunt force because football practice started tomorrow.

"Great," he muttered. After a pause, he continued, "I'm going to bed."

This last statement was directed at his folks, who were

comfortably seated in their usual living room spots. Sometime during a repeat episode of *Baretta*, his dad had dozed off in the recliner with the newspaper in his lap and his reading glasses halfway down his nose. His mom sat on the sofa peeling apples the boys had picked from the trees in their backyard. Peeling was the easy part. Helen Winters actually spent more time cutting out the worm damage, but the mother of three didn't seem to mind the tedious chore. The remaining fruit was halved and placed into a large, metal bowl resting at her feet. Homemade apple pie would highlight Sunday's dinner. Randy stood up from a second recliner and headed toward his room.

"Good night, Randy," said his mom.

"Night," Randy responded. He was halfway down the hall when his dad joined in the evening closure.

"Give 'em hell tomorrow," Paul Winters said. He understood how important football was to his son, and he was proud of Randy for putting in some extra work to prepare for the season.

Randy finished his walk down the hall and turned into the bedroom he shared with his younger brother, Mike. Tonight, Randy enjoyed the room to himself. Mike was staying at a friend's house for a slumber party. The house Randy shared with his family rested in unusual peace this evening because both Mike and Brian, the eldest of the Winters' boys, were away. Brian, now a high-school sophomore, was hanging out with the buds and celebrating the end of two-a-days, that grinding two-week period at the beginning of football when teams practice both in the morning and afternoon in preparation for the season.

Normally, Randy would have joined his friends, too, in some sort of evening adventure. After all, it wasn't only Friday but the last Friday before school began. However, tomorrow was a big day, the first day of practice for the Collins Junior High ninth grade football team. Randy had been waiting for this day all summer.

Compared to high school, the junior-high football program was modest. The season lasted two months, and the school budget only allowed for one coach per grade. Seventh and eighth graders didn't start practice until the week after school began. Ninth-grade teams in the Casper School District were allowed to practice beginning on the Saturday prior to school's start, and coaches could conduct two practices on that day.

After entering his room, Randy rotated a quarter turn as he pulled his T-shirt over his head. When the room became visible again, Randy stood face to face with his poster of *The Zonk*. Larry Csonka, number thirty-nine in his teal Dolphin's jersey and white pants, was pounding through defenders for another big gain. Randy revered Larry Csonka as his idol, and the Miami Dolphins were his team.

Since the seventies began, the Miami Dolphins dominated the NFL. They appeared in three consecutive Super Bowls, losing the first and then winning the next two. In 1972, they recorded a perfect season, a feat accomplished by only one other team. Led by Csonka, the running game keyed the Dolphins success. At 237 pounds, Csonka thrived as a monster of a man who chose to run over people rather than around them. Larry shared the backfield with Mercury Morris and Jim Kiick. Morris provided prolific speed while Kiick filled in as an all-purpose back. Kiick and Csonka were good buddies and received acclaim as *Butch Cassidy and the Sundance Kid*, characters from one of Randy's favorite movies.

Many of Randy's friends were surprised he hadn't chosen Kiick as his favorite player. After all, Kiick played college football at the University of Wyoming and led its 1967/68 Sugar Bowl team, the height of Wyoming's fame in football. But Csonka was always the man, and Randy devoured every magazine or newspaper article written about his hero.

Last year, the Dolphins again reached the playoffs despite several key injuries including one to Morris. On the first day of winter, the Dolphins traveled to Oakland and met the Raiders in a first-round playoff game. The Dolphins led 26 to 21 with less than thirty seconds remaining when Oakland's quarterback, Kenny Stabler, made history. That left-handed, lucky son of a bitch was falling to the ground when he threw a desperate pass downfield. No, he didn't even throw the ball; it was more like a shot put. Still, the ball sailed into the end zone where Clarence Davis miraculously came down with it in his hands despite being surrounded by three Dolphin defenders. The Raiders won 28 to 26.

Furious with the outcome, Randy bolted out the back door and climbed into the tree house Brian and he had built a few summers ago. This behavior constituted temporary insanity because of the brutal conditions outside. When the game ended at roughly 5:00, darkness had set in, the temperature registered thirty degrees, and the wind was gusting to forty miles per hour. Among the dormant tree branches, Randy brooded by himself for the next two hours until he was half frozen. Still miffed, he finally submitted to the cold and descended from his sulking spot only to learn the rest of the family ignored his tantrum and ate dinner without him. Mrs. Winters wasn't about to cave in to such foolishness. To add insult to injury, she let Mike be a total pig and gobble Randy's serving of spice cake.

Brian couldn't wait to begin the prodding, "Nice team you got there, Randy. Wonder if they're any good at golf. Guess we'll find out."

"Oh, bite me," Randy hissed.

"Can't find it. Don't have a microscope."

"Boys! Enough," scowled Mrs. Winters.

Randy continued, "They're a lot better than the *Raaaams*." He was still pissed. "The Rams," he muttered. "What a bunch of losers. They're gonna get smoked tomorrow."

Unfortunately for Randy and the Dolphins, the playoff loss wasn't the worst thing to happen during the winter. In 1974, a sports maverick hatched a goofy idea to create a second professional football league. The new entity known as the World Football League generated notoriety by pirating players from the NFL. While many teams lost some of their star players, no team was more affected than the Dolphins. A wealthy team owner from Canada signed the Dolphin trio of Csonka, Kiick, and wide receiver, Paul Warfield, to play in the new league starting in 1975 at an unheard of guaranteed contract worth three and one-half million dollars. In one severe blow, Miami had been reduced from pro football's greatest potential dynasty to an also ran.

Randy felt like he'd been punched in the stomach when he heard the news. He'd been betrayed by his sports idol. No! He was mad at Csonka, for sure, but who could really blame the other players or him for taking the money. It was those bastards running the WFL whom he despised. When the WFL started its season a few weeks ago, Randy refused to watch a single down even when the summer schedule offered no other worthwhile TV viewing.

Now as Randy stripped to his underwear, his attention returned to his own situation on the gridiron. Throughout his years in organized football, he had been cursed. Randy suffered the plight of a star running back living in an offensive lineman's body. At the beginning of each season, Randy jogged onto the practice field prepared to take handoffs and sprint around the end. Yet, every season he trudged off the same practice field lamenting the fact he was always lined up against some enormous bastard with a creepy giggle who liked bashing you in the head with his forearms. However, this season was going to be different.

Randy first sensed a change in his football fortunes during the final week of school last spring when Herr Potato announced he wouldn't be coaching the ninth-grade team. Herr Potato, what a name. Damn, Borias could be funny sometimes. Anyway, Herr Potato, formally known as Rolf Osterhagen, had been Randy's eighth-grade coach and was next in line to fill the role of the ninth-grade spot after Coach Dunkirk, a legend at Collins, decided to retire. However, Osterhagen's wife—Would she be Frau Potato? Good question. Check with Borias—was pregnant with the couple's second child and was due to deliver in early September. Osterhagen decided to take a one-year sabbatical from coaching, and that was great for Randy because

Herr Potato, although a decent guy in general, never appreciated Randy's talent as a running back.

Now Randy had been granted a fresh start, and he wasn't going to blow this opportunity. When summer vacation began, he created his own training plan which included pushups, sit ups, sprints, distance runs around the block, and footwork drills. Throughout the summer, he remained faithful to everything but the distance runs; he just hated that shit. Randy sat on his bed and studied his reflection in the mirror. He had grown another inch during the summer without gaining any weight. He felt stronger, faster, and tougher. He was ready.

Randy stood and crossed the room to turn off the lights. When he returned to his bed, he pulled back the blanket and bedspread leaving only a sheet because the night was so warm. As the back of his head absorbed the cool sensation of the pillowcase, Randy's thoughts strayed from football and onto other topics. He often used lights out as a time to sort out important matters and instill a sense of order in the world.

Of course, a teenage boy's world revolves around pop culture, and Randy considered himself an expert in all matters of practical importance including music and fashion. Further, Randy liked to make lists and rank things in order. He held mini tournaments in his mind about different types of trivia. As he clasped his hands behind his head, Randy recalled some of his previous events. There was the contest for the best food commercial where Randy quickly discovered these competitions were complicated and could be full of controversy.

Such was the case when he considered Beefaroni and Beefaghetti. In his mind, they were two separate foods, but both used the same commercial to advertise themselves as the most scrumptious food in the whole wide world. Randy was into *roni* but not so hot on *ghetti*, and while he would have liked to cut Beefaroni a break his duty to be an impartial arbiter of the rules trumped any personal favoritism. For competition purposes, the two foods were considered as one. Ultimately, it didn't matter because Oscar Meyer and its way with b-o-l-o-g-n-a smoked them. In the end, Fig Newtons won the title. Randy didn't especially like the cookies, but damn, you had to love the two-legged, big fig promoting his gooey inside and flaky outside.

Randy remembered his best superhero competition, and again controversy ensued. This time, the question centered on Johnny Quest and whether he qualified as a superhero. Had he qualified, the blond-haired boy would have skated away with the title, but, alas, in Randy's commitment to serve as an honorable judge he ruled Dr. Quest's son was a regular human being, albeit an ass-kicking, super-cool human being. The Quest verdict meant Randy now had to agonize between Spiderman and Space Ghost in the championship. After all angles were considered, Space Ghost simply commanded too much versatility with the power bands he wore around his wrists.

Randy wasn't especially sleepy, and his mind continued to wander. As his thoughts drifted back to the present, girls permeated his brain. Girls and thoughts about sex were becoming frequent topics during his lights-out time. Randy had yet to partake in sex, but adolescence fully controlled his body. Whether peeking at the cover of *Playboy* or watching his female classmates reveal some tempting thighs in their short skirts,

Randy found himself to be constantly locked and loaded, so to speak.

He knew it hadn't always been that way. True, girls injected an unusual fluttery feeling in him for as long as he could remember, but these sensations quickly passed. They didn't hold a candle to important stuff like pitching a tent in the backyard for a sleepover, playing with Tonka toys in the dirt lot down the street, or watching a western with your best friend at the downtown theater and then coming home and reenacting all the best fight scenes. Randy had only pondered his passage from childhood to teenager for a moment before the idea struck him.

"Ladies and gentlemen," the ringmaster operating in the center of Randy's brain began, "We are pleased to share for your pleasure, a nighttime listing topping all the other trivia madness Randy's ever considered. Here and now, may I present an all-star cast of women who gave Randy a boner before he knew what a boner was."

Ursula Andress of James Bond fame immediately popped into Randy's head, but strangely the voluptuous vision of her wading out of the ocean in snorkel gear wasn't the image that registered. Though *Dr. No* was the first film in the James Bond saga, and Randy was an enormous Bond fan, this version of Ursula didn't rate because he never saw *Dr. No* until it played on television one Sunday evening, years after its original release in theaters. By then, Randy was fully cognizant of erections.

Instead of great cinema, Ursula grabbed Randy's attention in that terrible movie, *Casino Royale*, starring David Niven and Peter Sellers. During one scene, she led Peter Sellers past a fish tank and engaged in all that lovey-dovey stuff on a sunken couch.

A song, *The Look of Love,* was playing in the background. Randy had heard the word *sexy* before but never really understood its meaning. As he listened to the music and watched this enchanting lady in a white feathery outfit seduce Sellers' clueless character, the mood shifted into something more mature than lovey dovey. It was sexy and one of the few decent parts of the movie.

Diana Rigg checked in as another Bond girl making the list. Not only did she look great in *On Her Majesty's Secret Service*, Randy's favorite Bond film, but she sizzled as a super hottie in television's *The Avengers.* Like *Casino Royale*, this show counted as another dorky waste of time. Randy wondered if the British were conspiring to torpedo television using beautiful women and ridiculous scripts. He made a mental note to research.

Randy recalled pop singers and their television appearances. He remembered Nancy Sinatra and her tall, white go-go boots; Petula Clark with that cute little mole on her jawline; and Bobbie Gentry, a sultry Mississippi queen with long, dark hair draping down over one shoulder. Why did Billy Joe jump off that bridge anyway? On TV, he loved to watch Jeannie, Samantha Stevens, Ginger, and especially Mary Ann.

As Randy cataloged his childhood crushes, his brain wasn't the only part of his body participating. His single bed now resembled a small pup tent. He continued to reminisce. When he thought of Katherine Ross and her role in *Butch Cassidy and the Sundance Kid*, his heart nearly stopped. He remembered the moments: Katherine Ross walking into her dark bedroom where a stranger awaited, Katherine Ross releasing a frightened and surprised cry, Katherine Ross unbuttoning a seemingly endless number of undergarments and exposing her cleavage—how many clothes do women wear anyway, Katherine Ross undoing the bun on her head and shaking her hair, shaking her hair, Katherine, hair, shaking. In Randy's opinion, watching the brown-eyed beauty with the pouty lips shaking her head of silky hair represented the quintessential boing-boing moment.

As if the fire in his groin needed any further fanning, Randy's hand slipped under the sheet. Visions of his favorite ladies streamed rapidly through his mind: Audra Barkley from *The Big Valley*—stroke, Agent Ninety-nine—stroke, Ann Margret—stroke stroke, Catwoman—stroke stroke stroke stroke.

Normally, the progression of male excitement completes its natural course ninety-eight percent of the time, but just before Randy achieved his pinnacle, an image of Space Ghost appeared out of nowhere.

"Must resist temptation," the superhero struggled to say. "Think negative thoughts."

"Negative thoughts. Okay, Space Ghost, I'll try," Randy confirmed. His mind whirred like a blender as he attempted to reverse the process. "Negative thoughts, let's see. How about *That Girl*, maybe Sister Bertrille?"

"No! Try harder," Space Ghost admonished.

Randy continued processing. "Edith Bunker? Mrs. Kravitz? Alice?"

"Alice, the Brady's housekeeper?" asked the superhero.

"Yeah."

"Much better," replied Space Ghost.

"Okay, keep going," said Randy in a self-encouraging manner. "Let's see," but he regressed, and his mind turned blonde with the image of Ellie Mae Clampett—stroke stroke stroke.

"My God, man! What are you doing?" screamed Space Ghost.

Randy reeled in desperation, but then lightning struck. "Granny," he cried.

"Yes!"

"Granny, Granny."

"Excellent!"

"Miss Hathaway!"

"Even better!" Space Ghost cheered.

"Granny and Miss Hathaway. Miss Hathaway and Granny," Randy bellowed triumphantly.

"You've done it, Randy! Your world is safe once more," the hooded cartoon figure congratulated him before disappearing.

Randy succumbed to exhaustion. He tilted his head back on the pillow and stared at the ceiling. After three deep breaths, Randy fell into slumber. Outside his window, the cottonwood trees swayed in concert with the wind and laughed at the antics of this silly boy.

PUUK pulled back from the monitor as Randy concluded his solo, man dance. This kid was wearing him out. The alien had originally been drawn to the boy's over-the-top worship of consumer products, an obvious irrational behavior. To understand the human's ravenous desire for sensory stimulation, the space traveler first shadowed the teenager via video and audio surveillance. Then he documented and later confiscated samples of the material items Randy mentioned. Now, a few shelves in PUUK's vessel resembled a small store. He had accumulated assorted edibles from Hostess and Frito Lay, footwear by Adidas, and a stack of magazines including *TV Guide, Sports Illustrated,* and *Penthouse.* His head was still aching from the combined, overwhelming scents of Hai Karate, Aqua Velva, and English Leather with which he had liberally slathered his body, and whose aromas now continued to drift in the craft. He decided hydration and rest would cure the discomfort, so he drank a pinkish fluid supplement and reclined.

At 6:45, Randy's clock radio clicked into action. Randy awoke to Glen Campbell singing *Rhinestone Cowboy* on KNAT—Natrona County's top-forty radio station. He hopped out of bed and began preparing for his own rodeo. After a quick trip to the can, Randy returned and shuffled through his dresser. He chose his lucky Dolphins T-shirt, navy shorts, and a pair of tube socks with navy and silver bands, the colors of the Collins Junior High Explorers. As he walked into the closet he shared with Mike and picked up his new white football cleats, a nervous knot formed in the boy's stomach. The day of reckoning had arrived. His football destiny would soon be revealed.

Randy headed to the kitchen for breakfast. After opening the pantry door, he pushed aside the box of chocolate fudge Pop Tarts, normally a staple in his morning regimen, and grabbed the box of Wheaties. He, indeed, wore his game face. His dad walked into the room just as Randy finished his breakfast of cereal, a banana, and orange juice.

"What time do you need to be at practice?" asked Mr. Winters.

"It starts at 8:00, but coach said to be there five minutes early," Randy replied.

"I'll give you a ride. We'll leave here at twenty 'til," his dad answered. Paul Winters managed the C.W. Furniture store. The C.W. stood for Central Wyoming. Saturday was a popular day for furniture shopping, which meant a workday for Paul. The store opened at 9:00 on Saturday, and normally Mr. Winters didn't head out of the house until 8:30, but he wanted to give Randy a ride, and he could use the extra time to catch up on some paperwork.

After breakfast, Randy packed up the rest of his gear and stepped outside to get in the brown, paneled station wagon. This morning felt different from those of the past few days. At an elevation of five thousand feet, daylight's first hour during late August often contains a slight chill, a harbinger that autumn is just around the corner. Yet on this day of clear blue skies, the temperature had already cleared sixty-five degrees. It felt more like a July Fourth morning. More important to Randy, it felt like a lucky morning.

As the boy and his dad drove to the junior high, Mr. Winters quizzed Randy about the day's activities. Was he coming home between practices? Had he brought a lunch? No, Randy would be going out with the guys for burgers or something. Did he remember his mouthpiece? Yes. When was the afternoon session finished? Practice was done at 4:00, but then the team had a meeting until 5:00.

Randy felt positive vibes as he entered the locker room. Being on the ninth-grade team meant he had been assigned a locker and didn't have to haul his football gear back and forth every day like he'd done the prior two years. He laid the small bag containing his jock and extra clothing on the bench and greeted the other guys who were present.

"Morning boys," Randy started. "Are you ready to see your new stud fullback in action?" He spoke in a louder tone than usual, as if he were trying to conceal something—nervousness perhaps.

"Only if Hell freezes over," replied Trev Baker. Trev wasn't a fan of Randy. The feeling was mutual.

A booming voice suddenly pierced the locker room. "Hey, listen up. After you put away your things, head on into the gym for a meeting. Take off your cleats before you go in. The floor's just been refinished, and we don't want it scuffed. Do not, I repeat, do not wear your cleats in the gym. If you do, you'll be running The Trail until you puke."

Not exactly Captain Kangaroo with his happy talk, Randy thought to himself. He removed the white cleats he'd just finished tying and noticed some of the guys were still wearing black spikes. *Not everyone's a trendsetter*, he concluded.

The team spent half of the morning session in the meeting. The boys were formally introduced to their new coach, Clyde Fullman, a product of Torrington, Wyoming. He earned all-state honors as both a linebacker and offensive end during his high school tenure. He attended a Division III college in the Midwest where he switched to defensive back. Unlike Herr Potato, Coach Fullman maintained a fit physique; he looked like he could step right back onto the football field and bust some skulls. Last year, he served as an assistant coach at Casper Senior. Senior is where the kids from Collins would be attending high school next year.

Most of the meeting covered routine, boring crud like reviewing the school code of conduct, practice times, practice vs. game uniforms, maintaining grades, blah, blah, blah. Near the end, Coach Fullman talked about the offensive and defensive sets they'd be using this season. Defense would be playing a fifty-two, monster alignment—the same scheme Senior was using. The offense was going to be a simple T formation although Coach Fullman said at times they would be sending one of the ends out into a split formation. They were going to run the football most of the time with a variety of power plays and counters. The day just kept getting better for Randy.

When the practice portion of the morning session finally commenced, the players weren't required to wear any equipment, not even their helmets. They spent the morning learning the calisthenics routine, jogging The Oregon Trail, and running forty-yard dashes for time. The practice wasn't overly taxing, but the day's heat left everyone sweating. Running The Oregon Trail was a bitch. The Trail, a hard-packed, dirt path, meandered through some barren land containing large deposits of alkali soil adjacent to the school's playing fields. The school grounds sat atop a small plateau so running the Trail meant a downward trek on the way out but an upward climb on the return journey with one pitch known as the gut sucker being especially steep. The gut sucker served as the grim reaper for the guys who failed to exercise during the summer break.

In football, the forty-yard dash—a sprint timed in tenths of a second—represents the definitive standard for determining speed. Here, Randy planned to claim his stake to a backfield position. Players ran in pairs, and every player ran the dash twice.

When Randy assumed the sprinter's stance for his first run, butterflies flitted against the pit of his stomach. At the sound of the gun, he exploded off the starting line and blew through the finish with furious strides. The timer's voice yelled, "5.4," and Randy burst into a winning grin because he had just bested his time from the previous year by 0.2 seconds. Randy's dedication to his summer training regimen fully paid off during his second attempt. His dash was clocked at 5.2 seconds.

Randy was elated with his performance. With his awesome times recorded, he decided to knock on the coaches' locker room door as the team filed into the school building after practice.

To Randy's surprise, Bernie Rickers answered the door instead of Coach Fullman. Rickers and Fullman became friends last year when they both served as assistant coaches at the high school. Rickers continued coaching there, but today, he was helping his friend out because the high-school team had the day off.

"Hey," said Coach Rickers when he opened the door.

"Is Coach Fullman here?"

"Yeah, come on in."

As Randy entered the medium-sized room, he noticed Coach Fullman standing over a desk and removing a chain from which a silver whistle dangled.

"Hey, Coach," Randy began.

Coach Fullman looked at Randy silently for a moment while his brain shuffled through the roster of player names he was trying to learn. In another second, his mind made a connection.

"It's Randy, right?"

"Yeah, Randy Winters."

"What's up?"

Brimming with confidence, Randy made his pitch. When it came to playing in the backfield this season, he wanted to leave no stone unturned. He intended to make sure the new coach and he were on the same page regarding his position.

"I just wanted you know you don't have to worry about fullback. I'm your guy. You can count on Randy Winters, number thirty-nine, to blast through the line." He spoke with resolute conviction.

As Coach Fullman took in Randy's presentation, he mentally noted some inconsistencies between the boy's and his own perceptions. First, jersey numbers hadn't been assigned, so he wasn't sure from where number thirty-nine came. Also, during a few conversations with Rolf Osterhagen, Randy's name had never been mentioned as a potential running back. Coach Fullman's observations during this morning's practice reinforced that notion.

He glanced over at Coach Rickers and then joked, "I'd love to Randy, but the problem is you run more like Shelley Winters."

Coach Rickers busted out in laughter as an image of the plump, middle-aged actress came to him. He remembered watching her swim underwater in *The Poseidon Adventure.*

With this picture in his brain, Rickers added, "How are you at holding your breath?"

Stung initially by the coach's remarks, Randy recovered and countered. "Hey, I just ran a 5.2 forty this morning," he stated flatly.

Coach Fullman answered, "Not bad, Randy, but there were at least ten players this morning who ran a 5.1, and three of them broke 5.0."

This fact stunned Randy. Sure, he knew he wasn't the fastest, but ten guys? He had been so preoccupied with his own performance he hadn't considered other guys were getting faster too. He was fighting a losing battle but was unwilling to capitulate.

"Yeah, but, but they don't have the moves or the power. They don't run like Larry Csonka," the flustered boy stammered.

The coach now understood the mystery of number thirty-nine, and he recognized the crushing disappointment in Randy's face. Clyde Fullman was not only a talented coach but also a person who knew how to manage people, especially young impressionable people. Rather than continuing to stomp on the boy's dream, he tried redirecting the conversation.

"Randy, you got great enthusiasm. I like that in my players. I can see you as a leader on the offensive line, maybe as the center calling out blocking schemes."

Center! Randy thought to himself. *Jesus Christ! Who wants to be a center? Nobody ever asked Larry Csonka to call out some goddamn set of blocking schemes.* He said nothing but could feel tears starting to well up in his eyes.

Fullman had been coaching long enough to know you can't make everyone a happy camper. "Go grab some lunch," he ordered, "and show me what kind of football player you are this afternoon."

"You gonna eat your curly fries?" Jerry asked.

A bunch of the guys had trekked three blocks to Arby's for lunch. They sat in groups of four in several booths. Randy dined with his good friend, Jerry Prentice; Mason (Rietzy) Rietz; and Phil Castinnori. Arby's rated highly as a Casper teenager's treat.

Although located near Collins Junior High, the fast-food chain normally failed to draw a midday crowd from the school because students weren't permitted to leave the grounds during lunch. Besides, thirty minutes wasn't enough time to get there and back. During summer break, most of the kids weren't willing to make the mile or more walk from their neighborhoods. So, on those few occasions when you could eat with your friends at Arby's, well, that was a special day. Yet, not even an Arby's ham and Swiss cheese sandwich paired with a Jamocha shake could raise Randy's spirits. He'd been sulking since he left the coach's locker room, and the entire team knew why.

"No," said Randy and pushed his tray toward Jerry.

"Do you think we're gonna have to run The Trail again this afternoon?" asked Phil. "I hate that damn thing."

"Do you know anyone who likes it?" Jerry replied.

The guys finished eating and walked back to the school. Upon entering the locker room, Randy passed by Trev Baker who couldn't resist digging at Randy's situation since the loud-mouth had been yapping all summer about how great he was going to be.

"Hike," mocked Trev while making the motion of a center.

"Eat me," Randy replied.

Joe Maldonado watched the exchange. Despite being the largest guy on the team, Joe moved well for a big man. As one of the best players, he would be starting on both offense and defense. Maldo, his nickname, was also one of those outgoing, happy types people loved being around. Maldo sauntered over to Randy, grabbed Randy's head in the crook of his arm, and pulled him close. Then he began singing the theme song from *Chico and the Man* reminding Randy not to get discouraged.

After Maldo repeated the verse, a dozen guys laughed including Randy. The disappointed boy currently didn't realize it, but spontaneous moments like these would carry him through the tough times. By ninth grade, football had turned into four parts work and only one part fun. Beginning now and going through the end of high school, players would have to endure increasing doses of pain, scream sessions, and conditioning, but sharing all the memories and the miseries with your buddies made it worthwhile.

As the players began to dress for the afternoon practice, a disgusted voice yelled from a corner of the room.

"Gross!" exclaimed Jerry. He had just stepped into his jock and cringed when its clammy dampness, complements of the morning practice, clutched his private parts. He felt even slimier when he pulled on his recycled T-shirt.

"Didn't you bring two sets of clothes?" asked the player seated next to him.

"What a dumb ass," contributed an anonymous voice.

Randy hadn't made Jerry's mistake. Having an older brother may not be an advantage all the time, but it often gives the younger sibling a heads up on potentially distressing moments. Randy dressed into his dry pair of athletic underclothing.

Unlike the morning practice, the team now walked onto the field in full uniform: helmet, shoulder pads, and pants. After calisthenics which now included shuttles of back pedals, side straddles, and crab walks, Coach Fullman separated the team into backs and linemen.

What had begun as such a promising day could now be described with a single word, sucky, completely sucky. Randy was sure the temperature exceeded ninety degrees, and here he was marching off to push the five-man sled with the big boys.

Coach Rickers commanded the drills with an unrelenting cadence.

"Hit it! Come on, drive with your legs. Hit it! Keep your butt down and head up. Hit it! You guys move like a bunch of old women. Hit it! Damn it. Get your butt down. Move fellas! Move! On the whistle, numb nuts. Hit it!"

When Randy's group completed their turn for the umpteenth time and rotated to the back of the line, his head throbbed from the tightness of his helmet. New equipment, courtesy of the energy boom in the Casper area, represented a double-edged sword. You were happy to receive it, but breaking it in was no fun. He wondered how things could possibly get any worse, and his question was soon answered when a swarm of flies appeared out of nowhere.

What the hell? Randy wondered silently. *We live in the wind capital of the world, and today when we need it there isn't a breath of air.* The flies continued to buzz around his face.

"Who shit their pants?" Randy complained absently but aloud as he positioned himself into a three-point stance.

"Hit it!"

After what seemed an eternity, the team stopped for a water break. Randy pulled off his helmet and immediately began rubbing his forehead, convinced he now sported a giant dent in his skull. Gradually, he walked across the field and past the chain-link fence separating the school building from the field.

Because he lingered, he found himself at the back of the line for the water fountain, but he didn't care. Randy was still massaging his forehead when he finally reached the water fountain. Because of the pain on his brain, he only half-noticed Rietzy, the only other player near him. Had Randy been paying attention, he would have honed in on his teammate's odd behavior. Randy finished his drink, reluctantly put on his helmet, and buckled his chinstrap.

"I did," Rietzy confessed as Randy started walking back on the field.

"You did what?" replied Randy in an uninterested fashion. He continued walking.

"I crapped my pants," Rietzy revealed in a desperate whisper. "I didn't mean to. I thought I was passing gas. Could you really smell it?" The poor guy was practically frantic.

Randy stopped and bit hard on his mouthpiece. Normally, he would have fallen on the ground rolling in hysterical laughter, but this bizarre incident just seemed to fit perfectly with his surreal day.

"No, I just said that because of the flies."

"Don't say anything, okay?"

"I won't. Come on, let's go."

Eventually, practice and the team meeting mercifully ended. A few minutes after 5:00, Randy shuffled out of the locker room and into the parking lot with Jerry. He spotted his dad's car, walked to it, and opened the front passenger door.

"Hey, Dad, can we give Jerry a ride home?"

"Sure thing," replied Mr. Winters. "Hi, Jerry."

"Hi, Mr. Winters," Jerry answered after he climbed into the back seat.

The boys were tired and silent, so Mr. Winters broke the ice after driving a few blocks.

"How was practice?"

In a matter-of-fact tone, Randy replied. "Well, Rietzy shit himself."

"No!" exclaimed Jerry in disbelief as he quickly leaned forward in his seat.

"Yes. Yes, he did," Randy assured his friend. "But don't tell anyone."

"I won't," promised Jerry, not really sure if he would or he wouldn't.

Shortly after the day's revelation, the station wagon pulled up to Jerry's house.

"Thanks for the ride, Mr. Winters. See ya, Randy."

"You're welcome," said Mr. Winters.

"Later," said Randy.

Randy and his dad continued home in a quiet car. Finally, Paul Winters asked the question to which he already knew the answer.

"So, does Collins have a bruising new fullback?"

"No," said Randy. He snapped his head to the side and looked out the passenger window.

Mr. Winters remained silent until they were just a few blocks from their home.

Finally, he said, "You know, I think your team's gonna have a helluva line this year. I think it's gonna be the cornerstone for the squad."

Visibly, Randy refused to acknowledge his dad's prediction, but after some extended thinking he agreed. They would have a damn good line, and the team should finish in the top three in the league and possibly could win it all. Come next week, he would play for the team, and he would play his best. Still, today had been an enormous disappointment. Today, his bubble had burst. Today, he would be sad.

That night, Randy wouldn't make lists, and he wouldn't dream dreams. He was pooped. Instead, when he laid his head on the pillow and closed his eyes, he was overwhelmed with the sounds, smells, and images recorded earlier in the day. He visualized the flies buzzing toward his face. He could feel sweat droplets stinging his eyes, but in his final moments of consciousness a particular set of sounds stood out the most. Randy heard the coach's shrill whistle and his gruff voice saying, "Hit it!"

PUUK

Where to start? That was the question the traveler pondered. Should he look forward to his destination and then back fit the gaping holes, or should he begin at his origination and move forward? Hmm. Maybe he should flip a coin. Flip a coin? Wasn't that a blatant example of the problem he was trying to solve? Did he really consider applying random selection over classic deduction?

These were unusual times. The attack on the monolith had been the latest incident in a chain reaction of bizarre behavior. The planetary inhabitants who existed for so long in a controlled coma of uninterrupted invariability had awoken in both a literal and figurative sense. The occurrences at his origination point were the impetus for his journey, so he should reasonably begin at the source of the issue—no, not the source but *The Source*.

The Source stood out as the only absolute in the universe. It was easy to understand, yet, impossible to explain. Scientifically advanced beings pinpointed its existence, but no one had determined exactly how it worked or when it began. *The Source* emanated bursts of low-intensity energy at a fixed rate. Each burst, formally known as a *ripple*, expanded as an

ever-widening circle. Because a ripple carried only a minis-
cule level of energy, it rarely affected any other celestial body;
however, it provided one invaluable service. Ripples served as a
constant measure of time throughout the universe.

The Source astounded the traveler's species because it
resided in a state so simple, yet, so powerful. The more they
failed in their attempts to understand its properties, the more
highly they regarded it. They adopted its uncluttered clarity as
the foundation for their own existence.

Though these alien creatures experienced no luck in
unlocking *The Source's* secrets, they realized tremendous success
in gathering information about other civilizations throughout
the universe. They mastered the techniques of travel in all four
dimensions. Fortunately for lesser beings, they refrained from
exercising their dominance and instead chose only to feed their
curious nature through non-invasive observation. During the
millions of ripples of exploration—for they were an ancient
population—life and death was witnessed in all its forms. The
observers were always saddened by tragedy but felt especially
despondent when they watched a world suffer from a self-in-
flicted calamity.

Being a meticulous population, the aliens collected vast
amounts of data regarding war and other types of intentional
destruction. From this information, they made two general
conclusions. First, combatants usually fought to gain more of
something. At the individual level, beings battled for food,
shelter, or mates. When the conflicting sides grew larger, stakes
included wealth, power, and influence. Second, fights often
took place because the involved parties perceived some sort of

difference between each other. Whatever the difference might be, at least one side believed the characteristic it possessed was in some way superior.

Although the traveler's race generally behaved as a congenial bunch, interaction with outside worlds influenced their attitude. Domestic turmoil began rising precipitously, and both the inhabitants and their home planet teetered on the verge of destruction. During the most desperate of days, the alien society found salvation in the simplicity of *The Source*. They constructed a non-descript monolith as a testimony to *The Source's* clarity and then, in their own language, inscribed an ideology to save their planet. Converted to English, the line read:

Consciousness And Conformity Are Cosmic Ideals

The ideology addressed both of the conditions deemed to ignite conflict. Typically, consciousness means the awareness of something, but under this new ideology it meant an understanding and awareness of everything. The consciousness movement, as it came to be known, emphasized the development of one's mental faculties while simultaneously downplaying physical rewards. Collective knowledge became the quest; grant each individual the capacity for complete understanding leaving him free from want.

The second term in the ideology was even more straightforward. If beings fight because they are different, then make everyone the same. Individuality was highly discouraged, but coaxing a given entity to renounce and surrender its unique essence is an exercise easier said than done. To persuade the population to comply, the *Thousand Mirrors* campaign was attempted. Basically, this public relations effort challenged individuals to stop looking inward and, instead, to fix their sights on their surrounding comrades. In doing so, they would observe a thousand identical reflections of themselves.

The *Thousand Mirrors* push only achieved moderate success. It was too serious. So in a peculiar twist, they resorted to a lighter approach. In what many would call an unexpected foreshadowing of their now current effort, the aliens remembered a catchy, feel-good musical pattern they heard during their inaugural visit to a remote planet called Earth. They borrowed a soft-drink jingle and changed the words to promote their own campaign.

I'm a conformer, he's a conformer, she's a conformer, we're all conformers, wouldn't ya like to be a conformer, too!

With great conviction and perseverance, the effort to ebb planetary discontent with these new values finally succeeded in all practical terms, and for nearly one-half million ripples tranquility ensued. The alien creatures perfected techniques in sharing a collective set of information, and each individual downplayed his uniqueness choosing instead to emulate his brethren. Over time, a virtual calm settled over the planet, almost as if its inhabitants had flat lined. They evolved into this steady state of consciousness minimizing the need for nutrition and activity. They almost lost all sense of feeling. Instead of being able to say, *I am hungry, I am happy,* or *I am frightened,* they were reduced to saying *I am.* The entire population actually stopped reproducing, and beings no longer died. They just sort of faded away.

However, despite their best efforts to conform into a perfect, single consciousness, they failed. Exceptions and variability started happening unpredictably in various forms. The purists, those beings possessing the highest levels of consciousness and conformity, were dismayed their mantra had become only a set of ideals and not an absolute; yet, as everyone realized, the only universal absolute was *The Source*.

Why couldn't absolute consciousness be achieved? Some said it was due to their inability to shield themselves from random contaminants. Though rigid precautions had been adopted to make their ecosystem impregnable, unexpected fluctuations in energy fields created brief openings and allowed foreign cosmic debris to enter. Others, however, believed the spark of life itself refused quenching. They thought all living things inherently strive to be something more than *I am*. Whatever the reason, odd behaviors by individuals continued to grow culminating in what the purists called the most grotesque of actions. Now, they had to deal with this infamous desecration.

To hear the purists' lamentations, an uninformed outsider would walk away thinking the end of the world is at hand. Yet, contrary to this impression, no sort of physical harm or damage, other than the newly carved line on the monolith, had happened. Instead, haphazard thoughts began interrupting the even stream of the collective's consciousness like hiccups. These mental spasms occurred in no predictable pattern, but they were juvenile and goofy in nature and fell into one of two types. They were either nonsensical or insulting.

Initially, the disruptions occurred in single-thought blips, but like an evolving organism they gradually grew into interactive, multi-thought incidents involving numerous sources. Having been dormant for so long, the collective's receptors dedicated to tracking exceptional mental activity required rebooting. When they were once again operational, the receptors recorded the first proof an assault on conformity was underway. The recorded event was called *Ein,* and translated into English contained the following set of mental bursts:

Guess what?
Chicken's butt.
That's stupid.
You're stupid.
You're more stupid.
And your brain's lopsided.

Similar episodes were captured with each incident containing varying degrees of malarkey. Yet, during just a few ripples, other puzzling changes took place. Audible speech and rudimentary physical movement staged a comeback with the latter centered round—of all things—a rediscovery of yo-yos. Individuals were once again talking aloud and learning how to maneuver their little toy *around the world.* Mostly, they chose to talk about yo-yos, and soon the whole conformity thing was blown away by yo-yo tricks.

Because they had been physically inactive for so long, only a quarter of the population possessed the fine motor dexterity necessary to successfully execute a *cat's cradle.* This variation in skill resulted in mass levels of whining. In time, sniveling carried over into other topics, and complaints quickly exceeded both nonsense and insults as the most common random outburst. Hence, the appearance:

But Bitchyness Is Universel

The traveler, for one, was actually pleased with the changes occurring on his home planet. Frankly, he thought, simply existing had become too dull for too long. He wasn't exactly sure why the changes were happening. He could accept random instances having some influence, but he also thought creatures should live as they were originally designed. In fact, he believed practically everyone was overthinking this dilemma. The traveler theorized the absence of reproductive activity, by itself, could explain a lot of the instability. Seriously, he would challenge any member of a species throughout the universe to go forty thousand ripples without getting laid and not be a little grumpy.

The purists jumped on the traveler's attitude as an example of contamination, and maybe it was. He wasn't alone in his attitude, but he clearly fell into the minority. Once the purists realized they couldn't prevent the *madness* as they phrased it, they looked for ways to control it. After some deliberation, they decided to approach the issue from a tolerance standpoint. Basically, they accepted the notion some random activity will inevitably occur and concentrated on determining what level of nonconformance was acceptable.

Using a sampling of random occurrences, a statistical model was devised to understand the level of nonconformity. The model incorporated both quantitative (e.g., the number of times an individual varied from expected behavior) and descriptive (e.g., what kind of random behavior was exhibited) measures. At first, the primary concern centered on how many times a given being was misfiring, but as nonconformance grew the degree of oddness in a behavior gained importance. For instance, crying out, *Wheee!* whenever a shooting star was observed could be tolerated in substantial numbers, but a single occurrence of someone drawing a moustache on the photo likeness of one of the purists was very disturbing.

The *Sigma System* was introduced to enforce an acceptable level of societal variability. While the goal remained for each individual to conform perfectly to the collective conscious, the *Sigma System* established tolerance levels for each type of unusual behavior. *One-sigma* actions were relatively frequent but harmless nonconforming behaviors observed in one-third of the population. Being dubbed a *one sigma* wasn't going to get a person in trouble, but it did exclude him from ever being designated a purist.

A *two-sigma* infraction indicated noncompliance with generally accepted uniform standards of behavior. Only ten percent of the population committed this type of misdeed and if caught would be subjected to corrective training. *Three-sigma* instances were rare and deemed very serious. They represented a behavior so bizarre it was observed in less than one percent of the population. Here is where the purists drew their line in the sand. These types of behaviors would not be tolerated, and banishment was being considered as a punishment.

Once the *Sigma System* had been established, the alien creatures concentrated on creating a method for tracking actual unconventional behaviors committed by the populace. Here, they quickly ran into *The Conformity Paradox*, a nasty little dilemma turning the conformity effort on its head. To reach the end goal of exact likeness among all beings within their race, they first had to recognize every individual as a unique entity. Therein lay the paradox. Basically, if odd behavior was to be minimized, each individual behavioral incident had to be assigned to the right someone so that someone could get fixed. For the assignment to occur, the capability to differentiate among individuals had to be in place—some type of unique identifier was needed. Crazy, huh?

Line two of the monolith's inscription appeared just as *The Conformity Paradox* was being recognized. The notorious action created an unprecedented sense of urgency for answers. Why had bitchiness become so pervasive on their planet? Moaning and groaning had been documented during their observation of other worlds, but it had always been regarded as an uninteresting, incidental item. However, when the constant bellyaching takes place in your own backyard, well, then it gets noticed in a big way. Who wants to share consciousness with a bunch of malcontents?

Because they knew complaining was a common characteristic on other planets, the alien creatures surmised those worlds might have insight on the condition and perhaps had developed some type of treatment. They reviewed all the historic data and determined the human population on the third planet revolving around Sol had exhibited an inordinate amount of griping along with other totally bizarre behavior. Consequently, they focused their initial attention there.

To refresh their understanding of the human culture, the alien creatures decided to address *The Conformity Paradox* using human tools. Specifically, they examined how humans uniquely identified one another. Given names were quickly ruled out as there were umpteen thousand people named Jose, Mary, or hundreds of other common monikers. However, they were encouraged by something called a social security number one particular country, the United States of America, had put into use. Essentially, the idea was to create a code consisting of enough different symbols so that every individual entity had a unique sequence of combined symbols.

The alien civilization maintained its own well-established language and symbols required for communication, but they were vastly different from those constructs used on Earth. If the alien creatures were going to gather data from Earth, they would have to be fluent in human symbols and languages. As practice, they decided to convert names for things in their world into a form humans could understand.

They chose English as the representative human language for a couple of reasons. First, it seemed to be understood, at least to some extent, throughout the entire planet. Second, it was the language used in the United States, home of the social security number. Upon deeper study, the U.S., an abbreviation the native population used, appeared to be a collection of humans from different geographic areas; by choosing the U.S., they could isolate their study of humans to one area and still maintain an excellent chance at observing groups of people originating from various parts of the world.

To begin, the alien creatures wondered how Earthlings might refer to them. In painstaking fashion, they translated their name for their planet into an English version. Symbol by symbol, the conversion was made. First came an *S*, followed by a *Y*, and then a *P*. The latter characters in their planetary name came across as multiple-letter combinations, one being *HIN* and the final one *CT*. They paused and considered the outcome of their efforts. Then, applying what they knew about English conventions for word extensions, they assigned themselves a name. The planet they called home was *Syphinct* making its inhabitants a bunch of *Syphincters*.

Next, they needed to assign a unique identifier to each individual member of the collective. Someone suggested, for fun, the Syphincters should create these identifiers using the English alphabet rather than the numerals used to create social security numbers. The purists gasped when the term *for fun* was mentioned because it inferred one-sigma behavior, but they approved of using letters rather than numbers. Instead of being for pleasure, the purists restated the effort's motivation as an exercise in perfecting their understanding and usage of English.

Just as the assignment of individual identifiers was about to begin, a minor crisis erupted on Syphinct. In an attempt to bring harmony back into play, a cooperative yo-yo rendezvous was convened. To the collective's horror, the event turned into a disaster and generated mass chaos. The event's participants had been divvyed up in groups of fifty-seven. Why fifty-seven? Well, it just so happens the Syphincters' term for tranquility translated into *Heinz*, and someone mistook one of the artifacts brought back from Earth on their previous visit as some sort of pacifying, liquid concoction.

Anyway, the groups of fifty-seven each formed a giant circle. In a demonstration of unity, every Syphincter was to simultaneously cast his yo-yo toward the center of the circle and then reel it back in. The release portion of the casting was okay but the distance and accuracy aspects were a complete mess. For most Syphincters, physical coordination was sorely lacking. Yo-yos quickly became entangled, and the reeling action cinched the multiple strands of string into tight knots. Planetary bitching reached its crescendo.

With one exception, every Syphincter was assigned to the untangling effort. The lone non-participant was tasked with creating unique identifiers and came to be known as the Administrator. Because the decision to use the English alphabet had been made prior to the yo-yo crisis, the Administrator's job was straightforward. The only piece of analysis still needed was determining how many letters would be used to make up each identifier. This question forced the Administrator to consider the total population of Syphinct, and he was surprised when he realized just how many of his companions had faded away during all those idle ripples. His race had dwindled to a mere 430,000.

Making a quick calculation, He determined every Syphincter could be uniquely identified using a four-character code, and so he began to make the assignments. The first randomly selected Syphincter would be known as AAAA; the second, AAAB; the third, AAAC; and so forth. No significance was attached to the assignment of one's identifier—at least that was the intent.

The Administrator was also charged with selecting someone to actually journey to Earth and perform the reconnaissance. The whole consciousness experience had extinguished the Syphincters' desire for interstellar travel; it was considered passe, so no one volunteered. One being would have to be randomly selected for the assignment—again, at least that was the intent.

However, the Administrator chose to take matters into

his own hands. Although his peers regarded him as a purist, a stealthy deviance dwelled in his inner core. He housed some internal feelings upon which he wanted to act. First, the Administrator suspected one particular Syphincter as being guilty of the heinous line-two inscription, call it a hunch even though hunches were clearly defined as one-sigma violations. His gut feel was reinforced while he was reviewing documentation collected from Earth. A human axiom had been recorded, and it struck a nerve within him. *It takes one to know one.* Though no one else had detected it, the Administrator was a deviant, and being a deviant gave him insight on spotting other deviants.

Accordingly, his suspect was assigned the duty of traveling to Earth. Of course, the clever Administrator disguised the selection to appear as if his suspect had been randomly chosen. He wanted to test a hypothesis. If the suspect were to leave the planet and the level of randomness were to drop during his absence, the Administrator could argue a circumstantial case the suspect was indeed the culprit.

As unique identifiers continued to be assigned, the Administrator watched with particular interest when the suspect's turn arrived. He was given the identifier, PVUK, and for a second time, the Administrator exercised his own will by changing the suspect's code. Again, he had been influenced by information collected from Earth, but mainly he was just being mean—a clear three-sigma infraction. During his reading, he learned a human term for regurgitation, a pretty gross concept in whatever galaxy you called home, was puke. To his own insidious delight, he swapped one character of the suspect's randomly assigned identity with another Syphincter, thus giving the traveler the undignified label of PUUK.

The Administrator savored his actions, yet, took great care to conceal them from the rest of the Syphincters. This feat was easily achieved because untangling yo-yo string was still everyone else's preoccupation. However, the Administrator should have been wiser and ultimately taken his knowledge of humans to a greater depth. Had he done so, he might have come upon another saying, *What goes around comes around*, but he didn't and continued operating in ignorant bliss as he formally recorded the random selection of PUUK as the designated traveler to Earth. He smugly notarized the official decree with his own unique identifier, PRIK.

To everyone's surprise, PUUK wasn't demonstrably upset about his selection as the designated traveler. Having been one of the first Syphincters to unbind his yo-yo, he secretly looked forward to experiencing different types of physical sensations. Of course, like PRIK, he had to subdue his feelings. He had already been tagged with a few two-sigma violations and didn't want to forfeit his good standing as a law-abiding Syphincter.

In preparation for his task, PUUK reviewed data collected from Earth so long ago. The amount of irrational material overwhelmed his mind. He extracted just a few examples:

Oedipus' choice of girlfriends—whoops

Fifteenth Century Europeans' belief the world was flat

The futility of Pickett's Charge at Gettysburg

Ralph Kramden's insistence on sending his wife to the moon despite a complete lack of astronaut training

PUUK inhaled a deep breath and then blew it out slowly. He had his work cut out.

Once the yo-yo disaster passed—gobs of string had to ultimately be cut and thrown away creating an insufferable bunch of whimpering—PRIK selected a small group of Syphincters to prepare PUUK's assignment. They assembled the following list of DOs and DON'Ts for his mission:

Limit observation to a preselected, fixed geographic area and time period.

Seek instances of odd behavior, and track subjects exhibiting those behaviors to identify any successful corrective action.

Observe individual subjects, but do not interfere in their lives. Exercise caution when interacting directly with any human subject.

Yada yada yada.

Upon receiving his instructions, PUUK readied his spacecraft. When all systems were go, he jettisoned away from Syphinct and toward days uncertain in another world. He was excited. Somewhere along the long journey, he retrieved the action plan for review. For the first time, he noticed an additional page affixed to the report. Quickly, he turned to it and read.

Stay out of saloons and casinos.

Bring back instructions for more yo-yo tricks.

PUUK couldn't help but sigh. His race really was in need of rescue.

JERRY

"I'm sick of everyone telling me what to do!" screamed the frantic female voice. "I want to make my own decisions, and I want to be left alone."

Today's ticket featured round three of the fight between Wendy Choate and her parents. To be completely accurate, Wendy was battling one parent and a significant other. She was arguing heatedly with her dad and his second wife, Pam Prentice. For the moment, both Wendy and her dad lived at Pam's house; part of the ongoing contention dealt with the rules by which the high-school junior was allowed to stay.

"You should have been home last night by 12:30," fumed Bobby Choate. "That is your curfew. Tiptoeing through the door at 2:45 with hickeys on your neck is bullshit. You're grounded for the next two weeks."

"We were very worried, Wendy," added Pam. "I'm disappointed you didn't bother to at least call and let us know you were okay."

Upon hearing the news of her grounding, Wendy blew into a rage. She pushed her chair away from the kitchen table, stood, and dashed to her bedroom, slamming the door behind her. A minute later, she reemerged—packed with defiance and an elevated set of vocal decibels.

"My curfew is only for weeknights!" Wendy screamed.

"Says who?" Pam retaliated. "That was *never* part of the deal."

The mercurial girl launched into Pam. "I can't believe what a hypocrite you are! You say you understand what I'm going through and want to be my friend. But then, you don't even trust me to make my own choices. I'm not a baby!"

"That's not fair, Wendy."

"Hey!" Bobby butted in. "No, you're not a baby, but you're only sixteen. And remember, Pam saved your ass when you had nowhere else to go."

Bobby Choate had participated in Wendy's second fight. After round one, a bitter argument between Wendy and her mom, Rita, Bobby sat in last spring when Wendy announced she would be traveling with Marvin, the beau, and some of his buddies to California for the summer. Bobby and his ex-wife responded with an emphatic no, but Wendy insisted she would. Since her parents divorced, Wendy had been receiving a significant monetary allowance from her grandparents, Rita's folks. She was counting on these funds for her journey; however, her grandparents sided with Rita and Bobby in this matter and threatened to cut off the allowance. Money talks.

As punishment, Wendy refused to speak to her mom or grandparents. Midway through summer, the situation between daughter and mom became unbearable. Coming to the rescue, Pam stepped in and graciously agreed to let Wendy live with Bobby and her.

A fourth person resided in the house, but after ten minutes of screaming, he escaped from the chaos. Jerry slipped out the side door, walked to the garage, and grabbed his bike. He needed some fresh air and peace.

"Man, what a bitch!" Jerry muttered. Life at home resembled anything but cheery since Wendy's arrival. He was never crazy about seeing her during the occasional family dinners they shared, but now with Wendy living under the same roof, he really began to despise her.

"People come and people go," his mom said. "I doubt Wendy will be here long. It's just a phase she's going through."

Jerry began walking his bike along the driveway, but the movement felt wrong. When he looked, he noticed a flattened back wheel.

Dang! When did that happen? He wondered. Jerry was capable of fixing a flat; he'd done so many times, but today he was already frustrated and didn't have the patience for the undertaking. He put the bike back into the garage and walked toward the street.

In his vessel, PUUK split his time between observing HUMAN SUBJECT: JERRY PRENTICE and cleaning his console. He hadn't anticipated the pressure packed into a canned, carbonated beverage and now was paying the price as he wiped off the sticky remnants of a Dad's Root Beer. Fortunately, his current subject was relatively inactive, so he could sanitize his workspace without sacrificing any research.

As he considered the teenager currently under scrutiny, he noted the lively interaction with the young female. In the alien's spontaneous picks for detailed study, he had failed to include any female candidates. PUUK hadn't yet determined if that was a mistake, but for the time being he would pay close attention to the effect this semi-related girl had on HUMAN SUBJECT: JERRY PRENTICE.

Meanwhile on Earth, the hot and still, mid-August afternoon played out under a canopy of clear, blue sky—a vintage summer's day in Wyoming. Jerry reached the public sidewalk, turned right, and started walking south toward Casper Mountain. At the corner, Jerry stopped and sat on the curb. A medium-sized crabapple tree provided sanctuary from the sunlight.

People come and people go. His mom's words continued to repeat in his mind. Wendy had been living with them about a month. It seemed much longer as every day was a struggle. Before Wendy, there was Bobby. Jerry's parents divorced when he was six. His memory of his dad was fading. He'd only seen him four times since his parent's split with the last occasion being a sixth-grade basketball game when Jerry fouled out. Last Jerry knew, his dad had moved to Alaska.

Bobby appeared on the scene when Jerry was in fifth grade. He earned his living as the head mechanic at Boulevard Motors, a Ford dealership. Jerry's mom also worked there as a cashier/accounts-receivable clerk. They dated for a while, and then last winter Pam and Bobby bolted one weekend for a Valentine's in Vegas jaunt. When they returned, Pam wore a wedding ring on her finger. Jerry's interaction with Bobby had progressed from *Say hello to my new friend Bobby*, to *Say good night to Bobby*, to *Pass the syrup and the funny pages to Bobby*. People come and people go.

Bobby was a good enough guy, but he wasn't a dad. He'd scoot over on the sofa so you could have a place to sit, but he wasn't going to enter your room to give you a pep talk when you had a bad day. As Jerry remembered, Bobby never hassled him, and his mom clearly loved having Bobby in her life.

Jerry placed his hands behind him on the grass, leaned back, and arched his spine. The stretching felt good. Like most Sunday afternoons, quietness prevailed. Jerry looked up Wrangler Street as it ascended the hill with Casper Mountain serving as a backdrop in the distance. Occasionally, a car passed across Wrangler on one of the four numbered streets that lie between where Jerry sat and the top of the hill. Jerry wondered how many times he had parked his butt on this corner and taken in the view, plenty for sure. Totem and he picked this curb as their talking spot. They would sit and eat Fudgsicles and shoot the shit about anything and everything. Pam always smiled when Jerry identified the corner as the boys' talking spot because neither of them were exactly conversationalists. In fact, most people struggled to extract a full sentence out of either of them.

Sometimes, Totem and Jerry would play a game by count-ing cars on 17th Street. One would pick the eastbound cars, and the other would take the westbound vehicles. They played to twenty-five, and the winner won a dime. Other times, they would pit streets against one another and tally the cars as they passed in either direction. 17th Street always won as it was one of the major roads in Casper. The suspense lay in the chase for second place; 15th and 18th Streets split almost evenly, but 16th Street never won, so sad.

The games Jerry and Totem shared weren't limited to counting cars. They both loved sports. Because Wrangler Street serviced only residential traffic, they could play catch in the middle of the road where cottonwoods, elms, and other tree varieties lined the street. They preferred to play in one of the yards, but it was impossible to find enough open space to throw the baseball in any of the grassy areas, so they made due with the hard surface. Unfortunately, the street badly scuffed the ball when one of them failed to make a catch, and any passed ball rolled along the pavement for at least two lots.

During the fall and early winter, the duo recruited some other guys and walked two blocks to Burnsley Park for some football. Most of the time, Jerry and Totem were limited to only younger guys as roster choices, so the pair would split up to make the game fair. Once in a while, they played together as one team against five smaller kids. The games were usually two-hand touch below the waist; but on those afternoons when a recent snow was melting, an unwritten rule dictated players must tackle each other and then wallow in the mud.

During the long summer evenings, up to a dozen neighborhood kids gathered to play cigarette tag, red rover, or twelve o'clock midnight. Squealing boys and girls navigated passages between houses trying to avoid capture. Though all the kids knew the obvious shortcuts, no one had mastered the neighborhood's topography like Totem and Jerry. The inseparable pair hopped fences, climbed atop garage roofs, and built tunnels in vacant lots. They were always the kings of the hill.

Jerry glanced across Wrangler Street and noticed the property on the other corner, the Jorgensen's place. Mr. and Mrs.

Jorgensen served as the senior members of the block, forever. Who knows how long they had lived there? They had three grown children with families of their own. One of their sons, Dane, lived in Minneapolis. Each summer he brought the family to visit. Dane and his wife had two children, a daughter named Sharon and a son named Robbie.

Robbie was two years younger than Jerry, but he hung out with the older boys whenever he visited. They let him pal around with them because he was a good guy and listening to him say the word *Min-ne-THO-ta* with his northern accent and slight lisp kept them highly entertained. In addition, he willingly shared his bitchin' collection of Hot Wheels.

With one exception, Sharon was a nonfactor in the boys' lives. One day when Jerry and Totem were playing catch in the street, Sharon took inventory of all the flowers growing in her grandparent's front yard. The Jorgensens had just returned from church, and she was wearing a short, frilly summer dress; shiny black shoes with a single strap; and socks that were folded over displaying a floral design. When she passed too closely to the rosebush, a couple of thorns caught a ruffle on her dress and gave the boys an extended glimpse of her panties, which were in a pattern identical to those of her socks.

During the next day's conversation at their talking spot, both boys timidly acknowledged a tingling excitement when Sharon's bottom was revealed. They also mistakenly concluded by looking at a girl's socks they could determine what type of underwear she was wearing.

The Jorgensens maintained a fabulous garden in their backyard. When Robbie was in town and with the grandparent's blessing, Totem and Jerry joined him in harvesting radishes, snap peas, and carrots. Some of the pickings were consumed on the spot, but most were carried into the house for Mrs. Jorgensen to use in her salad. When Robbie wasn't around, Jerry and Totem entered the garden clandestinely, headed straight for the rhubarb plants, and tore off a couple of stalks. Then they'd steal away in the alley, bite into the sour plants, and contort their faces with exaggerated puckers. One night, old man Jorgensen caught the boys in the act. As punishment, he made them hoe weeds and turn the soil over with spades for an entire weekend.

Peter Jorgensen was pushing at least seventy-five years of age. He had worked as a railroad engineer for almost fifty years, most of them based out of Casper. Even during retirement, his embedded Norwegian work ethic dictated his lifestyle. He always rose at dawn tending to his garden, his house, or his beloved 1958 Ford Thunderbird. On warm summer evenings, Mr. Jorgensen would sit outside his front door on the top step, smoke his pipe, and watch the world go by. He sat on one of those black welcome mats made with rows of small rubber prongs. Mr. Jorgensen had practiced this evening routine for so long, a butt print of worn nubs now existed where he sat.

Jerry wondered how many times Mr. Jorgensen had shared a bunch of grapes or a bag of peanuts whenever Jerry stopped to sit with him on those serene evenings. He also couldn't count the number of times the old man chewed his ass for cutting across his lawn with his bike or committing some other bad behavior. Last winter, Peter Jorgensen died instantly from a massive stroke. Mrs. Jorgensen continued to live in the house for the time being, but the place was rapidly deteriorating. People come and people go.

Jerry's thoughts drifted back into the present. Having leaned on his hands for several minutes, Jerry felt his palms going numb. He sat forward, rubbed his hands together, and then pushed himself back onto his feet. He didn't want to return to his house, so he began aimlessly walking toward the center of town.

In five minutes, he stood in front of Molly's Market, at the corner of 17th and Pearl. Molly's fell into that dying breed of Mom and Pop neighborhood groceries, but it wasn't so long ago when Molly's buzzed with activity. George and Molly Turner owned and operated the store. George insisted on calling the place Molly's as a tribute to his wife, whom he adored. The store consisted of three aisles in the middle for dry goods, a refrigerated/frozen area on one side, and the butcher section in the back. Two checkout counters were located at the front of the store, and delightful bins of assorted dime, nickel, and penny candies were displayed in front of the counters.

Yet, the best part of Molly's resided on the right side as you entered the store. First, it housed a number of intriguing vending machines including an old-style, red and white Coke machine dispensing eight-ounce glass bottles of pop for a dime. Besides a standard gumball machine, Molly's offered other novelty dispensaries. Jerry's favorite sold miniature football helmets of all twenty-six NFL teams. At a quarter apiece, Jerry calculated it would cost him $6.50 to collect replicas of the entire league. What hadn't been factored into his equation were duplicates. In fact, from September through November of 1971, Jerry dropped eight dollars' worth of quarters into the machine and still was missing seven teams. How can one person possibly get five helmets of the Buffalo Bills?

Further in from the vending machines, a small, ice cream freezer sat containing four flavors of hard ice cream. Vanilla, chocolate, and strawberry were staples, but the fourth flavor rotated from butter pecan to chocolate chip to rainbow sherbet. A single scoop on a small cone would set you back twenty cents, but you could add a second scoop for another dime. Clearly, a double was the way to go. The only downside to buying a cone was waiting until a clerk working the checkout counter was free to do the scooping; however, on hot summer days, an extra person was brought in solely to work the ice cream area. Even so, there were Sundays when a line extended outside the store's entrance.

A pinball machine stood in the back corner. A quarter bought you three balls and an opportunity to apply your flipper skills in an attempt to beat The Wizard. Players scored points by shooting the ball into a slingshot, hitting drop targets, and lighting up bumpers. Three thousand points earned you an extra ball, but the ultimate goal was to register the machine's high score because then you had not only beaten The Wizard but also every other player who'd paid his ante.

Unlike the bowling alley where older kids hung out in groups of five to ten and hoarded the two pinball machines, younger guys could actually get a turn at Molly's. Part of this opportunity stemmed from the Turner's conscious practice of preventing teenage crowds from massing. Everyone knew, at Molly's, you were welcome to buy an ice cream and play a game or two, but then it was time to move on.

Totem played pinball much better than Jerry. He wasn't the best, but he usually scored high enough to earn an extra ball. Still, Jerry remembered the Saturday Totem put up the machine's high score. It was a bit of a fluke, and it wasn't even Totem's best game. The machine had broken recently and when the serviceman came to repair it, he reset the machine's high score back to zero. The Wizard hadn't received much play since it had been fixed, and Totem took advantage by setting a new record with his first game and then bettering it during his third attempt.

Jerry distinctly remembered that day but not because of Totem's accomplishment. Instead, it was what Totem revealed on the walk to the market.

"We're moving," Totem announced.

"What!"

"Father found this great place just outside of town."

Jerry was a little stunned. "When did this happen?" he asked. The question revealed some hurt in his voice.

"He told us last night. I guess he's been looking at the place for a while."

Jerry recognized immediately he was about to lose something valuable, but he didn't comprehend the degree of that loss. He wasn't just losing his best friend; he was losing his way of life. The carefree days of childhood discovery had reached their end. Whatever was to come next, he would be facing it alone. People come and people go.

That had been two years ago. Totem's new place was really out in the sticks, so the boys found it impractical to visit each other often. Even a bike ride required significant effort.

Fortunately, the pair's time at Trooper Elementary was over, and Totem's new residence still fell within the boundaries of their new school, Collins Junior High. During the school year, the boys saw each other every weekday, and they were team-mates on the basketball squad. While they remained close, their friendship no longer matched the level they shared years earlier.

Likewise, Molly's best days were now behind it. A third national grocery chain moved into Casper, and Molly's simply couldn't compete with the pricing power of the larger stores. As Molly's loyal customer base died off or moved away, new residents shopped elsewhere. Backed by substantial television advertising, convenience stores also carved into Molly's rev-enues. Icee's and giant-sized fountain drinks became the new trend. Molly's ice-cream cone tradition remained strong, but it wasn't enough to keep the store going. The happy little place at 17th and Pearl would be closing at summer's end.

Jerry continued sauntering along 17th Street. As he headed toward downtown, every block contained a memory. He passed the public library where, as part of the Casper Summer Kids program, he remembered listening to stories about Wyoming and the Wild West. He never cared much for reading, but that expressive lady from the historic society with her reading glasses dangling on a chain could sure bring characters to life.

As he crossed Diamond Street, he looked down the block and spotted the doctor's office where his arm had been placed in a cast after breaking it. At age seven, he attempted to pick a piece of fruit just beyond his reach midway up the plum tree when he tumbled.

Two blocks later, Jerry stood in front of The Junction, a

popular casual restaurant consisting of booth-style seating. Customers placed their orders in each of the roughly thirty booths using a phone. The Junction was renowned for cheddar burgers with hickory sauce and heaping plates of onion rings. It had been their place, and today Jerry had passed by for the first time in weeks.

Jerry closed his eyes and pictured her. He saw and could even smell her shoulder-length, light-brown hair with straight bangs running across her forehead. Her roundish face featured a bright, warm smile and playful eyes; a smattering of freckles dotted her nose. He could hear her calling him, *Wimpy*. She used the name to tease him when he declined one of her challenges to push the limits whether it be putting super-hot sauce on a taco or skinny dipping after hours at a community pool. He usually countered by playing on the double meaning of her initials.

"Now BM, don't have a bm," he would tease back.

Jerry met Brenda Magruder at Collins in seventh grade. Four different elementary schools fed into the junior high, and Brenda arrived as part of the group from Prairie Elementary. She was a good friend of Tommy Barnes, and he met her through Tommy. Jerry and Tommy had played on the same little league baseball team for years. Along with Randy, who had come from Hillside Elementary, the quartet formed a mini social group within Mrs. Raymond's cluster of language arts classes. In seventh grade, students spent a total of three periods during the day with one teacher; one period was devoted to reading, another to English, and the final one was a pseudo study hall called English Work Period.

The group really banded together during the latter part of winter when the spelling team held practices after school. None of the four were exceptional spellers; Randy and Brenda were average while Tommy and Jerry stunk. However, Mrs. Raymond announced any person participating on the spelling team, a team she happened to sponsor, would be excused from taking the weekly spelling exams during the six-week duration of the season. Practices only lasted an hour on Tuesdays and Thursdays, and the opportunity to bag out on some classwork seemed too good to pass.

Brenda fit in naturally with the guys. She wasn't a tomboy, but she was clearly one of the buds. She enjoyed mixing it up with the boys and more than held her own. Over the course of the next year, they hung out often. Sometimes, the entire group gathered, and other times a splinter group of two or three assembled. Each member felt comfortable being paired with any of the others; yet, even early on, a special attraction formed between Brenda and Jerry.

The spark ignited the day after Easter during a matinee of *Benji*. Students enjoyed a four-day break for Easter; both Good Friday and the following Monday were days off. Tommy and Randy's families seized the break as an opportunity to visit out-of-town relatives, so the remaining two were left on their own. After a short phone conversation, Brenda and Jerry decided to meet that afternoon for a movie. The Rangeview Cinemas consisted of two theaters, and they were an integral part of the newest commercial complex in the booming oil town. *Benji* had been rereleased and was making its second run across the country.

They purchased their tickets separately and stopped at the concessions stand for drinks and popcorn. Brenda ordered Coke, and Jerry chose Sprite. Then they entered through the curtain and grabbed seats three rows from the back in the sparsely crowded theater. They laughed as the previews for *The Return of the Pink Panther* played. When the teasers for *Jaws* ran, Brenda nudged Jerry indicating this movie was a must see.

The feature began, and the audience watched the adventures of the star pooch. Sometime during one of the lulls between *aw* moments of little dog cuteness, Jerry accidently burped out loud. Brenda giggled, and Jerry delivered a side punch to her arm. As his arm retracted, his hand slid back and briefly came to rest on top of hers. Instinctively, Brenda's hand rotated so her palm faced up, and their fingers interlocked. Jerry drew a sudden breath and cautiously glanced at Brenda. She offered an accepting smile, and they both knew. They continued holding hands for the remainder of the movie.

As they walked out into the sunny but brisk March day after the movie, they were no longer holding hands. The couple turned left at the street front and continued to the first intersection.

"I told my dad I'd go to his office after the movie and catch a ride home with him," Brenda said.

Jerry nodded. He watched as the wind blew some strands of hair across her face. She pulled the hair back in place with her hand, and the pair studied each other in silence.

"So, does this mean we're going together?" Brenda finally asked.

Jerry paused briefly and then said, "Yeah, I guess it does."

Brenda smiled. Going together was a nebulous phrase dominating the teenage-courting experience. There were no formal procedures but many interpretations. Essentially, it was a public acknowledgement meaning *I like you, you like me, and we're okay if everyone else knows it.*

"Do you wanna come with me? My dad would give you a ride home."

"Nah, I'm good," Jerry replied while joining hands with her. Again, they gazed at each other in admiration.

"Well, I better go," Brenda said and started to release his hand.

Jerry pulled her close and kissed the unprepared girl. Brenda pulled back slightly. She scanned his face and then closed her eyes while tilting her head forward. As they kissed a second time, conflicting sensations swept over him. His heart raced furiously, but their lips shared a tender and controlled exchange. A sensual awakening intermingled with incredible vulnerability. It was the kiss of a lifetime.

Randy and Tommy picked up on the new situation immediately. The former was naturally perceptive and had been noticing subtle hints of amour for a while. He wasn't surprised. Tommy, however, was flattened by the development. He'd known Brenda for a long time and always thought she'd be *his* girl when the time came. Angry and hurt, he avoided the couple.

The final weeks of eighth grade gave way to summer vacation, and Brenda and Jerry were afflicted with puppy love. They saw each other almost every June and July day, partly because they wanted to—needed to they thought sometimes—and partly because they could. When it comes to staying busy, age fourteen is kind of a no-man's land. At fourteen, you're too young to get a real summer job but too old to go outside and play, so having someone you were crazy about to spend vacation with was really cool. In Jerry's case, it also provided an excuse to boogie from home and stay far from Hurricane Wendy. He thoroughly enjoyed being with Brenda and hadn't felt that kind of closeness to anyone since Totem had moved.

Typically, Jerry rode his bike over to Brenda's house, and events unfolded from there. Sometimes, they rode their bikes downtown or to the park. Almost every bike ride included a stop at The Junction even though they usually ordered no more than drinks. On other days, they hung out in the driveway shooting hoops with Brenda's little brother, Kevin. If it was during the weekend or in the evening, Mr. Magruder might come outside and join them in a game of *Horse*. He had played basketball in high school and possessed this wicked left-handed hook shot Jerry couldn't duplicate. Mr. Magruder laughed out loud and winked at his daughter each time Jerry missed and collected his *e*.

Other days were spent in the basement. The Magruders owned a young cat named Frisco, a little Siamese shit who would swat Jerry's arm as it dangled over the edge of the lounge chair when Jerry least expected it. While examining the scratches on his forearm, Jerry threatened to use Frisco as the basketball during the next game of *Horse*, and Brenda kicked him.

Jerry loved the basement and wished his home had one. It offered cool relief from the day's heat and sported both a TV and stereo. Best of all, it provided an outstanding place to smooch. However, too often, they had to share the area with Kevin and his friends. Even when they were alone, Mrs. Magruder demonstrated this uncanny knack for coming downstairs to water the plants or grab some supplies just before the good stuff was about to start. As June gave way to July, Brenda's mom began supplementing her basement appearances with poorly disguised suggestions it was time for Jerry to leave for the evening. That was annoying! July brought changes to Brenda, too.

Like mother like daughter, Jerry thought. He attributed the personality shifts to the heat.

Jerry couldn't put his finger on it, but Brenda seemed moodier, more serious. It wasn't related to that time of the month. No, she was different. Physically, she had definitely changed from their first meeting in seventh grade. Whenever they went to the swimming pool, he couldn't help but notice. Brenda had always been cute, but now she had this nicely-rounded butt and an ample set of breasts. She wasn't a little girl anymore. In shorts, she was flat-out smokin'.

Her body changes were the good part; the difficult aspect was trying to guess which Brenda was gonna show. Sometimes the carefree old Brenda appeared, but other times she angered easily and often seemed sad. Jerry was encountering a heaping dose of female complexities with their depth of character and ever-churning ideas, a no-win situation for a guy who spent ninety-nine percent of his time in the shallow end of the thought pool. The absence of any male mentoring on the subject further contributed to the strain on the relationship. Dad was long gone, and Bobby—well, it just wasn't going to happen.

"Jerry," Brenda began one day in an attempt to engage in conversation more meaningful than hollers for noogie time. "We never really seem to talk."

Jerry was silent. *Oh, oh. It's a mad/sad day*, he thought to himself.

She continued, "I mean, I don't really know much about you. In fact, I think I knew you better when we hung out at spelling club."

Jerry thought he had learned a little about calming girls. When it doubt, hug. He reached for her while uttering his best, "Aw, babe."

Brenda pushed his arms away. "You like to touch my boobs. I get that. But I'm really hoping I mean more to you than a quick feel. And I'd like to know more about you. What scares you? What do you see when you close your eyes? What do you think's going to happen to us? Things like that."

She paused and then finished. "I don't know, but sometimes when I'm with you I feel so alone."

Back in the present, Jerry stood only a block away from the center of town when he suddenly realized where he was and how far he had walked.

Wow, he thought. *Talk about a wandering mind.*

He glanced at the display window of Three Lizards Western Wear. A twelve by fifteen inch poster placed in the bottom left-hand corner advertised the Central Wyoming Fair and Rodeo. Casper's marquis summer event had now passed. The fair and rodeo took place each year at the July/August turn. Beginning right after the Fourth of July, Jerry and most of the kids in town counted the days until the fair.

For city kids, the fair boiled down to the parade and the carnival. Oh, sure, there were some other things to do. Exhibit halls displayed locally grown flowers, homemade jellies, and students' artwork, stuff that might interest the moms. Also, the rodeo was a blast, and it usually attracted two or three popular country and western performers. For farm and ranch families, the fair contained greater meaning. A season's worth of hard work culminated here, and kids in 4-H parted ways with the little lambs and calves they had raised.

Usually on a Wednesday, Fair Week kicked off with a parade. People crowded along sidewalks running adjacent to the parade route through downtown Casper. The event featured cowboys on horses, high school floats accompanied by marching bands, and the Shriners in their crazy little cars. Beauty queens and the governor rode in fancy cars and waved. When the parade ended, many of the visiting cowboys plodded into the bars ready to tie one on. Some people drove straight to the fairgrounds, but most either hung around downtown or headed home.

The carnival and rodeo ran Wednesday through Saturday. As a little boy, Jerry anticipated eating a corndog and cotton candy almost as much as he enjoyed the rides. In the last few years, the time spent hanging out with friends topped his list, and he thought that would be the case again this year. It wasn't. Jerry remembered that fateful Friday, probably the weirdest day of his life.

For days, Brenda and Jerry discussed their plans regarding the carnival. They would go together on Friday night. Brenda's family purchased rodeo tickets for Saturday, and Jerry promised Totem he would accompany him on Thursday. However, Brenda called Jerry with changes on late Friday morning.

"Hey, Jerry, tonight I think I'm going to the fair with my girlfriends."

"Huh?"

"It's Annie's birthday, and she wants to have sleepover afterward," Brenda explained. "And Tina's been riding my case about not seeing my friends all summer. You know, 'cause we've been together so much."

"Brenda, what the hell?"

"You got to go with Totem last night, right?"

"Yeah."

"Well, tonight it's girls' night."

"Fine. Bye!"

Two hours later, Brenda called again.

"Hey, Jerry, Tommy and some other guys are going tonight. He said he doesn't hate you anymore. Go with them, and meet us there."

At 7:30 that evening, the two groups met at the Zipper, and Jerry rode in the cage with Brent Mendenhall. At 7:45, the bunch lined up at the Skydiver. Brenda crept alongside Jerry and nudged him with her shoulder.

"You think you can handle this, Wimpy?"

"I think so."

Brenda and Jerry hopped into the caged car. Their car had barely elevated one spot on the giant wheel to allow the next car to load when Brenda flung her arms around Jerry's neck and locked her lips on his. For the duration of the ride, Jerry enthusiastically accommodated Brenda's passion while keeping one hand on the steering wheel.

At 8:30, Jerry sipped his Sprite. He noticed Brenda

whispering to Tina and pointing toward him. He walked over to the girls while smiling at Tina.

"Is she telling you what a good kisser I am?" he bragged.

Brenda glared at him and quickly walked away.

"Jerk!" Tina said.

At 9:00, Brenda rode the Scrambler with Tina and Annie. Jerry watched from outside the protective railing. At 9:30, having received the silent treatment for an hour, Jerry approached Brenda as they entered the line for the Himalaya.

"Ride with me. Please!"

She agreed. They moved into their seats and fastened the safety belts. The train of cars started forward in a circle with increasing speed. Midway through the ride, the cars stopped and then began accelerating backward. Jerry tuned into the music accompanying the ride. The James Gang was delivering some serious guitar licks.

"Man, I love this song," he said.

"Jerry, we need to break up."

When the backward revolutions finished, Brenda stepped quickly out of her seat and hurried down the Midway with her face in her hand. Tina and Annie raced to catch up. Jerry sat motionless. He listened one last time to the song's chorus and recognized the unexpected irony of the song's title, *Walk Away*.

More than three weeks passed. He called her a few times, but she wouldn't come to the phone. Mrs. Magruder always made some flimsy excuse. August had been awful. Not only did he miss her, but he was also bored silly. He thought he'd go crazy if he had to spend any more time locked up with Psycho Wendy in their little house, the one with no basement.

Jerry had been away from his house for quite some time. Still, he wasn't in a hurry to return. He decided to detour a few blocks out of the way through City Park. As he entered the park, his outlook brightened.

Football starts in a week, he thought. *and then classes.* He wasn't crazy about school, but it sure would be an improvement over spending more time at home.

City Park contained long rows of dense pine and fir trees running parallel to its walking paths. The trees provided a natural barrier between the half of the park with the band shelter and picnic tables and the other half with its open, grassy area. At the end of one of the tree rows, Jerry blindly turned and nearly ran into a couple. It was Brenda and a guy; they were laughing and holding hands.

Jerry quickly recognized the guy as Jim Schmidt. He was in high school, a junior or maybe a senior. Jerry couldn't remember. A shocked expression swept over Brenda's face but then disappeared as quickly.

"Hi, Jerry," she said evenly and then bit her lower lip.

"Hey, man," Jim added.

Jerry weakly exhaled, "Hey."

Only a moment passed, but it felt like hours of awkward, tortured silence.

Finally, Jim said, "You better watch out for those trees. They're dangerous." He laughed and then pulled Brenda past Jerry. "C'mon."

They walked across the grass toward the parking lot. Meanwhile, a voice from behind Jerry rang out.

"Hey, dip shit. What're you doing?"

Jerry spun around. Randy was walking toward him and grinning. His friend's arrival distracted Jerry from what had just gone down.

"Not much, what are you doing?"

"I just finished mowing a lawn, earning some bread. You know," Randy explained.

Jerry looked at Randy and then at the area around them.

"Aren't you missing something?" Jerry asked.

"What?"

"A mower?"

Randy chuckled. "Oh yeah, Brian picked it up. He had a yard to do." Randy partnered with his older brother in a lawn mowing business.

In the distance, they heard a car door shut. The boys looked up and saw Jim walking from the passenger side of a '67, burnt-orange Mercury Cougar to the driver's side. They watched as he started the engine, backed up, and drove out of the parking area. Brenda never looked their way. People come and people go.

"Did you know?" Randy asked continuing to watch the car.

"No," Jerry replied. Then while staring down at his feet he added, "It don't matter."

It mattered—a lot. Jerry was inundated with different feelings, most of them bad. Then it dawned on him what happened; Brenda had traded up. Everyone knows girls like older guys, and this older guy drove a hot ride. He was probably a better kisser, too. Jerry obviously couldn't confirm that, but he had no doubt Brenda could.

Randy watched his friend mope for a few moments and

then said, "Hey, let's go over to my house. I'll beat your ass in ping pong."

The boys exited the park on the opposite side from where Jerry had entered and headed for Randy's place. Randy spotted a medium-sized rock lying in the gutter and booted it in the direction they were walking. When they reached the stone again, Jerry sent it skimming with his foot. The boys took turns kicking the featureless rock down a desolate street as afternoon faded into evening.

JIMBO

The slender body sporting tan thighs in cutoff jeans and a beautiful head of blonde hair rounded the downtown corner. Within seconds, the wolf whistles began.

"Baby, baby," an admirer yelled. "Come out and play!"

A second voice cooed, "Ooooh, looking fine. Mmmmmm."

Totem and Jerry stepped onto Main Street remaining a few paces behind the source of attention. Having witnessed similar occurrences previously, the boys just shook their heads. Meanwhile, the object of affection revealed a smile that was wide, white, and electric.

All the hoots and banters emanated from a red, four-door Chevy. Its passengers consisted of females in their late teens or early twenties. As the auto trolled past, one of the back seat riders leaned out and shouted.

"What's your name, little boy?"

His name was Jimbo DuPree, and he was simply gorgeous. He threw a friendly wave at the passing auto and waited for his compadres to catch up.

"Jesus, Jimbo. How much did you pay them to raise a fuss?" Jerry teased.

Jimbo shook his head and laughed. "Just doin' what comes naturally," he stated.

And he was. His easy-going nature combined with head-turning looks made him virtually irresistible. Further, his delightful southern mannerisms played favorably with most new acquaintances in Casper. This central Wyoming city was his family's third home in five years. They moved to Wyoming a year earlier from Texas via Lafayette, Louisiana. Jimbo DuPree mixed Texas pride with Louisiana craziness. His was a Mardi-Gras demeanor spoken in a Lone-Star drawl.

Jimbo accidently ran into Totem and Jerry downtown. The latter pair was bumming around on this Saturday afternoon; Jimbo had arrived to get his hair cut—no, not cut but styled. To Totem's way of thinking, the difference between the two terms amounted to eight dollars.

"Seriously, you paid seventeen dollars for a haircut?" Totem asked again as Jerry and he rejoined Jimbo. The transaction weighed heavily on Totem's mind since Jimbo had mentioned it ten minutes ago.

"I got it styled, man," Jimbo explained. "It's not just a cut. First, she uses a razor blade on my hair to do some of the trimming, and then she puts some conditioner in and massages my scalp. It feels so good. Ooh!" A little shoulder shimmy accompanied Jimbo's exclamation.

"She?" Totem asked in surprise.

"Yeah, a lady."

"You pay a lady seventeen dollars to cut your hair?"

"Pretty much. Plus tip," Jimbo said in a matter-of-fact tone.

Totem shook his head. Jerry now intervened and changed the subject.

"So are we headin' to The Mod Shop or Three Lizards?" he asked.

After the boys unexpectedly met, they decided to visit some clothing stores together. Jeans ranked as the top consideration. The Mod Shop offered a hip, trendy inventory aimed at fashion-conscious youth. It carried the latest designs such as men's shirts in bold prints and pants that flared out below the knee. Three Lizards—a Casper institution in western apparel—sold traditional straight-legged jeans, both Wrangler and Levi's. It was also regarded as the go-to place for flannel shirts, belts, and boots.

"Mod Shop," Jimbo answered, and Totem shrugged in agreement.

PUUK, who was monitoring the boys' activity from his vessel, booed loudly at the decision. During his brief surveillance, the space alien had already acquired a great fondness for western wear. He gave his highest admiration to garments covering the top and bottom of the human body, cowboy boots and hats. He studied his own one-piece suit in its drab gray color and dreamed of alternative attire. In the not-so-distant future, PUUK promised himself a fancier set of duds.

Back on Earth, the trio walked two blocks and then descended the small flight of stairs leading into the store. Denim was enjoying a particular surge in fashion, and this season it was offered in multiple shades. The boys paid special homage to a mannequin dressed in faded denim jeans with a matching jacket.

"Those are some righteous threads," Totem admired.

"I eee," Jimbo agreed.

I eee was Jimbo's manner of saying yes. When people heard this response from Jimbo for the first time, they thought he was being a smart ass, but it didn't take long to realize it was just part of Jimbo's personality. It was unique to him in the same way *Affirmative, Captain* was unique to Mr. Spock. Again, Jimbo was just doin' what came naturally.

Although their eyes were big, their wallets were light. Accordingly, the boys recorded mental notes of clothing they would like to purchase later or perhaps receive as a gift during Christmas or their birthday. After leaving the store, they headed toward All Star Sporting Goods. Jimbo revisited a topic they'd discussed when Jimbo first ran into the guys.

"So, tell me again how you've got a car," he requested. "You ain't sixteen yet, are you?" The question was directed at Totem.

Totem began, "Well, you see I've got a special kind of driver's license. It's called a hardship license, and I got it because we live outside of town. There's no bus, so I don't have no way to get to school unless I drive."

"How long you had it?"

"I got it in April."

"I never saw you drive to school last year," Jimbo said.

"I didn't," replied Totem now shifting his posture into an uncomfortable stance. "Edward, my brother, used to take me." There was a long pause. "But now he don't live with us too much anymore."

Jimbo wasn't familiar with Edward. In fact, he didn't know Totem or Jerry very well. He had met them through Randy at school. They all ate at the same lunch table, and he'd shared a few classes with each of them. On the other hand, Jerry knew Edward quite well as the Sweetgrass family used to live on Jerry's street prior to moving out in the country.

"Edward's been doing some time this summer in the slammer—selling weed," Jerry offered. Jerry's explanation fell short in accuracy. Edward was, indeed, selling weed, but he hadn't been busted for it, at least not yet. Edward's recent arrest dealt with drinking.

Totem clenched his jaw, and his body tightened. Jimbo nodded silently in understanding. Jerry tried to lighten the mood.

"So, with his brother gone, Totem gets Edward's car."

"It's William's car!" Totem snapped with his eyes burning.

Dang! I just stepped in it big time, Jerry thought to himself. Aloud he said, "Sorry man, William's car."

Jimbo didn't know the specifics, but he caught the drift. He waited a few seconds for the emotions to calm and then slowly broke into his signature smile. Looking directly at Totem, he said, "Well, hot damn, man. Those are some nice wheels you get to drive."

It wasn't Jimbo's words but instead his sincere delivery that put Totem at ease. He was smooth. He asked Totem some more about the car and learned Totem's license only permitted him to drive during the day. But come October, Totem would be sixteen and have a regular driver's license. The guys made it about halfway to the sporting goods store when Jimbo abruptly stopped. He motioned for Totem and Jerry to look across the street.

"Check it out," Jimbo said.

Three honeys were lingering around a cement planter across the avenue. Two were sitting on the planter, and the third stood facing the others. She was the one who caught Jimbo's attention. Like Jimbo, she was wearing cutoffs revealing shapely legs. She wore a white halter that tied behind the neck and a straw cowboy hat atop her head. The girl sitting on the left wore a cowboy hat, too, but it was brushed in felt and tan in color. The other seated girl wore a baseball cap. It carried some type of logo on its front, but from across the street the guys couldn't discern the lettering. She was sipping a drink with a straw.

Jimbo waited for two cars to pass and then crossed the black asphalt. Totem and Jerry followed like puppies. The girl with the drink noticed them first. She lowered the paper cup from her mouth back down to her lap and made a preliminary assessment:

Lead Guy—Absolutely adorable but young
Indian—Tall, probably quiet, mysterious
Other Guy—Tall, nice shoulders

"Hey there," Jimbo opened as he reached the curb. "What're y'all doin'?"

The standing girl turned to face them. "Howdy," she said looking at Jimbo. The attraction was immediate.

The girl in the felt hat applied a different tact and fired a shot across the bow. "You boys lose your mom?" she joked as she stood.

She looked slightly older than her friends. She wore a white T-shirt with a picture of Waylon Jennings in concert and had packed her full-sized butt into a pair of Lee Jeans. Her height was medium, and she was chesty. WAYLON was printed in large, navy letters over his picture. On her, his name stretched from—well, you know. Her verbal jab drew vastly different reactions from the boys.

Jerry became defensive and blurted, "Mom! Hell!"

Jimbo rolled with it, smiled, and replied with Texas ease, "Yes, ma'am, we did. Now we're just lookin' for love and kindness wherever we can find it." He knew how to play and continued.

"I'm Jimbo, and this here's Jerry and Totem."

"You're so full of it," laughed the Waylon fan."

"How old are you?" Jerry asked still miffed by the insult.

"Sixteen and a half," the girl in the straw hat stepped in and offered.

Jerry silently considered her reply. *Sixteen and a half*, he thought to himself. *Your body says twenty-two, and you talk like you're twelve. I guess that's about right.*

"I'm Patsy," she announced. She continued by motioning to Waylon. "She's Kim, and over there's Terri."

For the first time, Terri spoke. Looking at Totem, she said, "I seen you before. You beat us in basketball. My brother said your team played dirty."

Totem looked at her and slid one step to the left so he could frame her outline between Jerry and Jimbo. He noticed the logo on her hat was for Caterpillar.

"What school?" he asked.

"Glenrock," she replied.

"I remember that game," Totem said. "You had one guy who went nuts from the outside. He musta had twenty points."

"That's my brother."

Totem thought about the game. He didn't think they'd played dirty. They played rough. There's a difference. Both Totem and Jerry played for Collin's Junior High. When he wasn't in trouble with the coach, Totem started. His soft hands and height advantage met the requisite skills of a prototypical low-post player. Jerry also received plenty of playing time. He was a banger; even though he didn't start, he still fouled out about half the time.

"I bet you're a cheerleader," Jimbo complemented Terri. "I can just see you in your cute little outfit cheering your boys on." This illustration drew a rise from Terri.

"Yeah, I'm a cheerleader," she confessed, "but I don't cheer for those little turds. I'm in high school."

"The big time," Patsy added with pride. She was a cheerleader, too.

Kim interrupted. "It's almost 3:00. We better head back."

"Well, maybe we can get together sometime," Jimbo suggested. "Do y'all have a phone number?"

"Are you planning to ride your trikes twenty-five miles?" Kim zinged.

Jimbo looked confused. Glenrock, a small town of a few thousand people, was located about twenty-five miles due east of Casper. Being new to Wyoming, he hadn't yet mastered the regional geography. He thought Glenrock was just the name of one of the local schools.

"I got a car," Totem said, saving Jimbo the embarrassment of admitting his lack of understanding.

"Well, maybe we'll see you boys down our way sometime," Patsy said. She pulled a pen and a napkin from her purse.

"Here's my number," she said and handed it to Jimbo with a flirtatious smile.

"Maybe you will," Jimbo smiled back.

The Glenrock ladies departed leaving the boys to assess what had just transpired.

"Man, we just talked some high school chicks into giving us their phone number!" an elated Totem crowed.

Jimbo corrected him. "Actually, I just landed the phone number. But, you did rescue me on the Glenrock bit, so I guess we're even." Then he added, "I think Terri's interested in you."

"Yeah?"

Totem liked the idea. He was a rookie when it came to girls. Jerry carried an entirely different impression about the encounter.

"Can you believe that other girl, Kim?" he said. "Man, what a cocky wench."

"What do you care?" Jimbo asked rhetorically. "You've already got a girlfriend."

Reality's pain slapped Jerry in the face.

"Not anymore."

"You're not going with Brenda!" Jimbo cried with surprise. "Man, she's sweet."

Jerry couldn't have agreed more. "It got weird," was all he said.

Jimbo recognized the look of a fresh wound, so he didn't press. He turned to Totem.

"Think you can give me a ride home?"

"Sure."

The next morning Jimbo was seated at the kitchen table

with his family. They had just returned from church services and were enjoying breakfast. Mrs. DuPree, an extraordinary cook, had prepared scrambled eggs, homemade biscuits with sausage gravy, and cantaloupe for the group. Excited chatter filled the room

"I can't wait to see Yogi Bear at Jellystone Park," Eva chirped. She was one of Jimbo's five-year-old twin sisters.

"Eileen, please be careful. You're spilling gravy on your dress," Mrs. DuPree scolded the other twin. Then she lovingly corrected Eva. "It's called Yellowstone Park, baby. Jellystone is a pretend name in the cartoon."

"I just love Boo Boo," Eileen sighed dreamily while gazing up at the ceiling.

"Yeah," Eva agreed. Then she added, "Jimbo sez Ranger Smith is a dick."

Mrs. DuPree nearly spit out her coffee. "Eva, that's a terrible word to say!" She pierced Jimbo with a scathing stare and then regained her composure. "Park rangers are good people. They help take care of the forests and the animals who live there."

Her mom's reaction surprised Eva. She hadn't meant to say anything bad, but she felt safe because it seemed like her mom was blaming Jimbo for whatever was making her angry. Eva recalled the day Jimbo made the statement and remembered what else had happened, so she decided to share.

"I 'member when Jimbo said that 'cause it was the afternoon we ate chocolate ice cream. Jimbo said we wouldn't get caught if we washed the bowls and put them away before you got home."

Eva's pride in successfully pulling off the stealth snacking radiated on her face. "And you never found out. We fooled you, Mama!"

Laurie DuPree wasn't impressed. "And I bet that was the same day no one ate their supper," she countered. Eva's smile disappeared.

"Maybe Jimbo's a dick," Eileen suggested.

"Didn't I just say what an awful word that is!" fumed Mrs. DuPree. "The next person who uses that word is going to get the whipping of her life." She glanced over at Jimbo and added, "Or his."

The two other diners had stayed silent. Georgia and Earl Mason, a.k.a. Grams and Grandpa, were visiting from Texas. Grams disapproved of the twins' word usage and wished Laurie would have condemned the behavior more sternly. She was convinced the foul language was Bo's influence. Conversely, Grandpa chuckled softly and was thoroughly amused by the incident. Grandkids were the best. After Jimbo was tagged with the great ice cream caper, Grandpa shot him a sly wink.

The Masons traveled by car to visit their daughter and her family. They arrived late on Friday and spent most of Saturday resting from their journey. The second leg of their trip would start tomorrow when they would drive to Yellowstone Park with their three grandchildren. Meanwhile, Laurie would accompany Bo on a company retreat to San Francisco. It had been a while since they had spent any time as just a couple, and she was looking forward to the break.

Bo DuPree missed church and was also absent from the breakfast table. Skipping services wasn't uncommon, but he usually manned the head seat at the morning meal. While

Laurie was a devout Baptist, Bo could be best classified as an occasional Methodist. The difference in spiritual conviction caused some friction between them; however, as one of many compromises constituting a marriage, a happy medium was determined. Religious education for the children would fall under Laurie's domain, Bo would be excused from regular weekly attendance at church services, but he was expected to join the family in worship on Christmas, Easter, and other significant events such as a child's choir performance. Bo was actually planning to accompany the extended family this morning but was interrupted by business.

He was trained as petroleum engineer and now served as a field supervisor for Wahoo Technologies, a medium-sized drilling operator. Bo currently directed the drilling activity taking place in a developing oil field about fifty miles northeast of Casper. After an initial dry well, the company struck oil with a deep find at about seventeen thousand feet during its second attempt. The field held significant promise, and efforts to drill three more wells were underway.

Bo possessed a great understanding of both geology and drilling technology, but as supervisor he found himself spending most of his time managing the various players in the drilling business. Sometimes that meant keeping his roughnecks out of barroom brawls and jail cells. Other instances required calls and, on occasion, personal visits to state and federal officials who regulated the drilling operation process. During those occasions, Bo would plead with them for timely approval of a well permit. With the current, furious pace of drilling activity and the painful cost of letting a leased drilling rig sit idle—as much as fifteen thousand dollars per day—this particular activity consumed much of Bo's week, and he disliked it immensely.

On this Sunday morning, Bo had received a frantic, post-sunrise call from a dirt contractor whose road-building activity into a new well site halted abruptly after an irate rancher fired a rifle shot into the blade of a moving bulldozer. Apparently, during the clearing operation across the rancher's property, the bulldozer destroyed a thriving sapling growing near the edge of a creek bed. Landowners placed tremendous value on trees of any sort because of their scarcity in this part of the country. The little sapling was a favorite of the rancher, and he specifically requested it remain undisturbed. Bo remembered the original conversation vividly, but somehow the dirt contractor didn't *get* the message, and now guns were a blazin'.

Money or more of it can settle many disputes, but as he drove out to speak with the rancher Bo knew this infraction was personal. The best case, he thought, would be at least a week of downtime for things to cool off. Worst case would be the rancher's unwillingness to ever allow access across his property. The drilling site resided on neighboring federal land, and alternative routes existed. However, the next best alternative doubled the length of road and introduced a steep ravine into the construction equation.

Not surprisingly, when Bo returned to the house later that morning, his mood was sour. Yet, the magic of baby-girl hugs works wonders, and when Eileen wrapped her arms around one of her daddy's knees, his attitude softened.

After her embrace, Eileen stepped back and asked, "Daddy, won't you come with us to see the bears?"

Bo picked her up with his large hands and wheeled her around. "Not this time, Critter," he replied using his pet name for the twins. "You get to have fun all week with Grams and Grandpa."

It was Eva's turn to chime in. "And we get to wear our new dresses for dinner one night," she exclaimed while hopping up and down.

Bo smiled at his daughter's enthusiasm. Then he noticed his son who had been sitting quietly at the table.

"What about you, Jimbo?" he asked. "You'll need to pack a dress, too. Maybe a pretty pink one."

It was an odd statement, the type of thing one guy might say to another in a joking manner. Only this time, neither Bo nor Jimbo were smiling. Earl picked up on it but said nothing.

Laurie caught it too. "Bo!" she barked condensing *knock it off* into a single word.

Jimbo briefly made eye contact with his dad, pushed away from the table, and silently walked out of the kitchen.

Next morning, the group packed up for the trip. A small cold front had pushed through Casper overnight leaving a thin layer of clouds hanging in the east. They blocked the sun as the five travelers prepared for a long day's ride. The trip to Yellowstone would lead them west at first and then north to northwest as the day continued. Their plan was to stay in Cody for the first night, a trip covering 215 miles.

Along with some magazines and her purse, Grams slid into the middle of the white, Lincoln Continental's back seat. From years of experience, she knew keeping siblings apart on road trips meant keeping the peace. It was her own version of divide and conquer. Eileen and Eva took their spots on either side of her.

Jimbo hopped into the front seat with Earl. The visit to Yellowstone would be a first for everyone. Jimbo felt mixed emotions about this trip. He dreaded being cooped up with the twins for almost a week but loved spending time with his grandparents. They seemed to enjoy life so much. He was also relieved to slip out from under his dad's shadow.

Few people in the world know about the desolate, two-hour drive from Casper, west, to a little place called Shoshoni. Jimbo snorted at the first disruption on U.S. Highway 20/26, a wide spot in the road called Natrona. Its welcome sign indicated a population of ten. The car passed through Powder River, population forty-four, without anyone appreciating they had just seen the largest town between Casper and Shoshoni. About this time, the first wave of boredom swept over the girls.

"Grandpa," Eileen said, "can we listen to the radio?"

Grandpa obliged and found a country-and-western station that stayed in range for about twenty minutes. During this portion of the drive, the group passed Hells Half Acre, a small section of badlands some overly optimistic entrepreneur marketed as a tourist attraction. Jimbo adjusted the radio dial and tuned into KNAT; it broadcast a much stronger signal than its country-and-western counterpart.

When Led Zeppelin's *Immigrant Song* came on with its howling Nordic screams, Grandpa interpreted the noise as some type of car trouble and pulled to the side of the road. In between gut-busting laughs, Jimbo explained the odd sounds were part of the song, and Grandpa immediately shut off the radio.

"Damn rock 'n roll," he cursed.

The passengers motored along without any music through Waltman, population six; Hiland, population ten; and Moneta, population seven. Anyone who's made this drive can attest there are varying degrees of nothing, and the closest one approaches to absolute nothing happens just before entering Shoshoni.

Grams had realized relative success at keeping the twins occupied by leafing through her flower magazines. She carefully pointed out her favorite flowers from home. She loved bluebells and bluebonnets, honeysuckle and wild azaleas.

"What kind of flowers does Wyoming have, Grams?" Eva asked.

At that moment, Georgia Mason lifted her head and surveyed the vast, barren expanse. With a kind of dumbfounded confession, she admitted, "I'm not sure anything grows here."

But just as night is darkest right before dawn, the wondrous Wyoming landscape was about to become much more interesting. After leaving the metropolis of Shoshoni, population 649, Boysen Reservoir appeared immediately like an oasis in this arid land. The manmade body of water provided vital irrigation for farmers. From the reservoir, the water flowed down the Wind River revealing a spectacle millions of years in the making; the river had carved a twenty-four hundred foot deep canyon through the Owl Creek Mountains.

A drive through the fourteen-mile Wind River Canyon on U.S. 20 presents a geologic odyssey through Earth's entire history. Here, the great pressures that formed the mountains also buckled them. In doing so, they exposed rock from every geologic period beginning with the Precambrian and moving forward to the present. As the travelers entered the canyon from the south, the layers of rock had been uplifted and tilted downward. The canyon walls currently visible on either side dated back almost three billion years. When they exited the canyon, bright red stone from the Triassic Period was featured, a mere two hundred million years in age. Jimbo enjoyed the aesthetics of the canyon scenery, but only Earl appreciated the magnitude of the geologic wonder.

Soon after the Lincoln left the canyon, both front-seat passengers noticed a brown road sign.

"Wedding of the Waters," Jimbo read. "I wonder what that means?"

Upon hearing the word, *wedding*, ears perked in the back seat.

"Is someone getting married?" Grams asked.

"Can we go and eat some cake?" Eva cried stretching her neck to catch a glimpse of any sort of refreshment gathering.

Grandpa shook his head.

"I don't see any cars, and nobody gets married on a Monday morning. Besides, we're going someplace better. Should be there in a few more minutes."

Wedding of the Waters wasn't an event but rather a location. It is the point at which the Wind River becomes the Bighorn River. Geographically speaking, there is actually no change in the river. When white explorers entered the new western lands, separate groups had assigned different names to the waterway at different locations not realizing it was the same river.

Ultimately, government officials decided the upstream part of the river would continue to be called the Wind River, and the downstream portion that eventually drains into the Yellowstone River would maintain its identity as the Bighorn River.

The *someplace better* Grandpa referred to was the small town of Thermopolis. Although Yellowstone Park is world renowned for all its geothermal attractions, the world's largest mineral hot springs is located in Thermopolis, roughly one hundred thirty miles southeast of the national park. A few commercial spas were built around the hot springs, and Earl surprised the grandkids by taking them to one with a swimming pool and a gigantic water slide. The twins would always remember this place as the tepee with the hot water and funny smell.

After spending nearly three hours in Thermopolis playing in the pool and eating lunch, the group once again climbed in the car to finish the day's travel to Cody. After he had driven a few miles out of town, Grandpa slowly pulled to the side of the road.

"Your turn to take the wheel, Jimbo"

Jimbo was stunned. "Uh, Grandpa, I don't have my learner's permit yet."

Grams agreed. "Earl, what do you think you're doing?"

"The boy can drive. He done it in Rooster last year on the property."

The Masons lived on the outskirts of Kilgore, Texas. Grams grew some vegetables in her garden, Grandpa kept a few horses, and an odd assortment of chickens roamed about the place, but it would be a stretch to call the property a farm. However, it did lie outside the city limits enabling Earl to take certain liberties such as letting his underage grandson drive his red, '59 Dodge pickup, a.k.a. Rooster, out in the fields.

"That's different, Earl," Georgia protested. "He and you could get in big trouble."

"Do you believe in the boy?" Earl asked his wife point blank.

"It's not about that," Grams responded.

Giggling, Eileen added her two cents. "Daddy sez Jimbo don't know his bottom from a hole in the ground."

Earl ignored the child and addressed his wife again. "Do you believe in the boy?"

Exasperated, Georgia gave in. "Fine, fine!" she said. "But *you* will be the one to tell Laurie!"

"Slide on over, son," Grandpa said to Jimbo with a reassuring smile.

Jimbo moved over into the driver's seat while Grandpa opened his door and walked around to the passenger side.

"Nice and slow, boy. You'll be fine."

Part of Jimbo wanted to shit a brick, but another part felt good, real good. Grandpa believed in him. That was a nice feeling; one he hadn't known for a long time. Jimbo drove for eighty miles without incident. Five miles outside of Cody, Grandpa instructed him to pull over, and they exchanged places.

After spending a morning viewing artifacts from the Wild West and artwork by Frederic Remington and Charles Russell at the Buffalo Bill Museum, Jimbo and his family entered Yellowstone Park from the east for three days of touring. Jimbo was amazed at almost every turn. The scenery varied from dense, alpine forests to wide-open river valleys and from large, pristine lakes to steep canyons. The twins were less impressed.

Because they were young, their attention spans were short, and because they were small their eye level fell below the bottom of the side windows meaning all they could usually see was the sky and some occasional treetops.

Whenever the twins could exit the car and actually view an attraction, they appreciated the experience much more. Eva particularly enjoyed drinking lemonade in the grand hall of the Old Faithful Lodge and then watching the world's most famous geyser erupt. Eileen preferred navigating the winding boardwalks to reach the bubbling, multi-colored paint pots. Both disdained the sulfurous odors and jointly protested by holding their noses.

Prior to the trip, Jimbo already knew about the geothermal aspects of Yellowstone Park, but the Grand Canyon of the Yellowstone's raw beauty caught him completely by surprise. He watched the Yellowstone River's churning waters violently tumble hundreds of feet over both the upper and lower falls. The canyon walls in their hues of yellow, orange, and brown projected a prehistoric canvas that signified their forever past.

Traffic halted on their return from the canyon back to the lodge. Several bison were crossing the road and created an idling line of vehicles eight deep.

"My goodness! Aren't they a shabby bunch!" Grams exclaimed as the massive animals lumbered back into the tall grass.

Suddenly, two yahoos jumped out from the car in front of Earl's and with cameras in hand started to approach the bison. Earl glanced at their car's license plate.

"California," he muttered. "It figures."

As the reckless pair closed within thirty feet of the creatures, Earl began yelling.

"Hey, you jackasses! Move back! Those are wild animals. They're unpredictable."

The dual photographers fired a *screw-you* response Earl's way but hastened their camera clicks and went back to the car. Earl turned his head toward the back seat to address his granddaughters.

"Girls, I never want you to go out with boys like that. They're dumber than a fencepost."

With her widened eyes, Eva acknowledged she would heed Earl's advice, but Eileen wasn't persuaded.

"Oh, Grandpa," she answered. "Mama sez boys are just stupid. If I have to wait for a smart one, I'll never get married." Grams cackled uncontrollably at the five-year-old's profound reasoning.

When the chain of automobiles started moving again, the family drove another mile and then turned into a parking lot for a picnic area.

"I need to stretch my legs," Grandpa announced.

Jimbo and his sisters hopped out and ran toward the gurgling sound made by nearby water. The twins squealed with glee as they skipped along the banks of the Firehole River and watched cutthroat trout dart out from under the rocks and through the water.

They didn't see Yogi Bear or any other bear for that matter. A few years earlier, park management launched a concerted effort to discourage motorists from feeding the bears, a practice harmful to both humans and bears. The efforts were working, and bear sightings had decreased significantly. Although the family was disappointed, they enjoyed an unexpected delight when, with the assistance of binoculars, they watched an otter family frolic in a Hayden Valley stream.

Their trip wasn't finished, but when the family headed south, leaving Yellowstone Park in the rearview mirror, Jimbo was certain all the great sites had been seen. He couldn't have been more mistaken. At one point along the John D. Rockefeller Memorial Parkway, the Lincoln turned west and into a meadow clearing. Rising from the valley floor in the distance, the Tetons appeared. As these ragged, granite giants came into view, Grandpa stopped talking in midsentence, pulled the car over to a wide spot in the road and simply gazed at the blue wonders while the car idled.

Jimbo stepped out of the car and stared in silent awe at the magnificent peaks. He was overcome by a strange sensation. It took a few moments, but he finally understood how this stunning example of nature's majesty made him feel; he felt *insignificant*, but it was insignificant in a comforting sort of way. He imagined the mountains speaking to him.

Dude, your problems, whatever they are, really don't mean squat in the grand scheme of things.

His family checked into the Jackson Lake Lodge about an hour later. From their table in the formal dining room, the quintet basked in a gorgeous view of Jackson Lake in the foreground with the Tetons as a backdrop. Jimbo and his grandparents devoured the fresh trout, lightly fried. Eva opted for macaroni and cheese, and Eileen ate a hot dog. After dinner, Grams was pooped and ready for some down time, but she agreed to accompany the twins to the gift shop before they retired for the night.

The guys weren't ready to shut down yet and decided to spend the remaining forty-five minutes of daylight walking along a path near the lake. About one-half mile from the lodge, they turned onto a smaller trail leading into a forest of evergreen trees. For the first time on the trip, Jimbo found himself isolated with only Grandpa as company. Until now, all the touring had been a whole-group affair.

Initially, they engaged in small talk—the fullness of their bellies from dinner and the sudden drop in evening temperature. Then Earl changed the subject.

"Are you looking forward to gettin' back home and goin' to school?"

Jimbo thought about it for a moment. "Kind of," he started.

"Any big plans?" Grandpa asked.

"Not really." Then, partially ashamed, Jimbo admitted, "Grandpa, I'm not playing football this year."

Earl accepted the news with silent neutrality.

"Dad's really pissed," the boy continued and then immediately started to apologize for his choice of words. Earl waved it away.

***

In his day, Bo DuPree excelled in football for his high school in San Angelo, Texas. His school played in a prestigious, 4A west Texas grouping dubbed the *Little Southwest Conference*, a conference that included three-time state champion, Abilene High School. Bo played on both offense and defense, and he won accolades as a fleet-footed halfback. He earned second team all-conference honors during both his junior and senior year and was talented enough to be pursued by several schools including SMU, Baylor, and Rice. Ultimately, he chose to become a Bear.

During his first year, Bo made the traveling team but didn't play much. As a sophomore, he began the season as a second-string back but gained a starting role during the third game of the season when the player ahead of him broke his collarbone. Bo played for two games until that fateful October Saturday when a linebacker submarined him and destroyed his left knee. The hit ended his career.

Bo was reasonably smart, so the end of football didn't mean the end of college. Propelled by a strong work ethic, he slogged through a difficult engineering curriculum. He met Laurie at Waco during their freshman year but didn't begin dating her until they were juniors. They married five weeks after graduation, and Laurie gave birth to Jimbo a year later.

Bo loved having a son and wore his fatherhood proudly. As Jimbo grew, Bo reveled in teaching his son how to swing a bat or catch a pass. He encouraged Jimbo to participate in anything remotely related to sports. He envisioned his boy continuing the family tradition. As a youngster, the early training from his father gave Jimbo an advantage over the other kids his age. Bo was elated to see Jimbo batting third in the lineup.

Unfortunately, Jimbo's advantage over his peers diminished as he grew. Bo viewed this change as temporary; he knew kids grew in bursts rather than a straight line and was confident Jimbo would eventually make a name for himself on the diamond, gridiron, track, or possibly, all three. Yet, despite their common bloodline, father and son were two different people. Jimbo simply wasn't blessed with Bo's athleticism, and more important, he didn't have his drive. Bo stood out as a hard-working fighter; Jimbo was an easy-going lover.

Bo's vicarious expectations withered in disappointment at a time when the man had lost control over much of his own life. The twins, born nine years after Jimbo, were an unexpected twist. He adored them, but they had put Laurie through a difficult pregnancy, and what little down time he had previously known vanished. Then there was work and the ever-increasing stress to drill more wells with tighter deadlines. Now, his son had decided to become a quitter.

***

"I'm not that good," Jimbo admitted as he continued confessing to his grandpa. "I doubt I would even start."

Then, he remembered how much he shivered on the sidelines during those October practices when the wind whipped across the field and bit through the thin layers of his pants and jersey fabric. Jimbo hated his first Wyoming winter. By Thanksgiving, he was already looking forward to spring. Little did he know, the unrelenting winds and brutal temperatures were just starting.

"And it's freezing up here during football season. It's not like back home," the boy continued.

Earl considered his grandson's dilemma. "Have you found some other activity you'd rather do?"

Jimbo didn't get a chance to answer. They had been walking absently along the path. It led them from a section of lodgepole pines into a small grove of aspen. Having cleared the aspen, the path descended, and they now entered a wetland area populated with willow bushes. As they cleared one of the larger willows, a pond became visible. Roughly eighty yards away and standing in the middle of the pond, a bull moose casually looked up to assess the newcomers. Green moss was hanging off his snout. Satisfied the humans weren't an immediate danger, he dipped his face back into the still water and continued to feed.

The impressionable teenager, once again, was swept away by nature's grandeur. Jimbo had faithfully attended church his entire life, but until this trip he'd never felt touched by God.

RANKIN

Rankin Bellamy was a homosexual. Oh, he wasn't really a homosexual, but he was a little faggot. What's the difference?

The difference is stark. A homosexual is an adult who sexually or romantically partners with a person of the same gender. A faggot, or fag for short, carries a few meanings. First, like the word, queer, it can be a slang expression for homosexual, or it can be used as a label to identify a guy who behaves in a soft, effeminate, or quirky manner. In the younger crowd, the term is universally employed as a putdown.

Rankin Bellamy wasn't a homosexual—at least not yet and probably would never be. First, he was not yet an adult. Society was debating homosexuality and whether sexual preference was a voluntary choice by an individual or a prewired condition in a person's very being. Whether a person chose or was destined to be homosexual, he at least recognized who he was as he matured into an adult. Rankin hadn't yet reached that point in his life, so one could only speculate about the boy's outcome.

However, PUUK would soon abduct Rankin and collect relevant information about his persona. The data would indicate with a high degree of confidence the boy would be straight. During this abduction, Rankin would be subjected to a variety of physical and mental examinations from which the alien would conclude, along with tons of other conclusions, the boy was attracted to the opposite sex. PUUK would predict the boy's sexual future in jargon both clinical and dull. In human terms, Rankin could be expected to follow the manly mainstream. He would soon hatch out as a teenage horndog, morph into a lustful American male, and ultimately decay into a dirty old man.

No, Rankin Bellamy wasn't a homosexual, but he was a little faggot. He was a little faggot because some of the guys at Collins Junior High called him a little faggot. Junior high is a special time because it is a time of change. Teenagers start their junior high years as children and then transition into becoming adults, but while many changes take place, the types of changes occur at different rates of speed. For many boys, in particular, physical maturity and an independent attitude tend to happen long before intellectual and emotional maturity materialize. By the age of twenty, most aspects of human maturity have been completed, but in those early teen years—those junior high years—the gap between physical and other types of maturity peaks.

Perhaps more than any time during a person's life, junior high passes as a time of the haves and have nots in which the former preys upon the latter. Big guys dominate the small, reckless fellows intimidate the cautious, and the aggressive feast on the passive. In 1975, Rankin was small, cautious, and passive. He possessed all the makings of a little faggot.

That's not to say Rankin was a loser—no, far from it. He had many things going for him. He displayed both smarts and musical talent. Rankin wasn't as gifted as his older sister, Regan, in the musical realm, but he had played piano for eight years, and with a little practice he could master just about any musical piece asked of him. Rankin might have levered his ivory-pounding skills to advance his popularity had it not been for his mother.

Constance Bellamy, a self-professed aficionado of treble clefs and quarter notes, ran roughshod over Rankin's choices in listening and playing. Modern music meant The Carpenters, not Alice Cooper. Classic American consisting of Gershwin, Bernstein, or Rodgers & Hammerstein offered an acceptable alternative to the European masters, but this cacophony of garbage now infiltrating Broadway with titles like *Hair* was simply unacceptable. Mrs. Bellamy even questioned the legitimacy of *Jesus Christ Superstar* and reluctantly approved only because Andrew Lloyd Webber had been formally trained, and she appreciated a few of the songs from *Joseph and the Amazing Technicolor Dreamcoat*.

Rankin's easy-going personality enabled him to cope with his mother's strict demands, and he actually enjoyed his mom's musical tastes. Likewise, he listened with content to whatever his peers would play when he was away from home. While he sometimes wished for his own preferences, he never caused a fuss. Rankin simply went with the flow, and in doing so he missed an opportunity to score some admiration points from his classmates. Rock 'n roll signified rebellion; although he could have easily jammed with any of the three or four garage bands playing in Casper, Rankin never swaggered with that independent, defiant attitude. He limited his public musical engagements to a few piano recitals and school band performances as a flute player.

Interestingly, Rankin held a special affection for one particular type of music neither his mother nor his peers approved. It was bubblegum, that genre of pop music defined by a catchy up-tempo beat and silly-sweet lyrics. Rankin watched *The Archies* faithfully on Saturday mornings to hear this sound. His mother considered bubblegum to be a waste of time and wondered why he couldn't follow in Regan's shoes and get serious about music. Still, compared to the other modern-day trash, she regarded *The Archies* as harmless. She let his appetite for bubblegum slide because while he listened to it, he was still a little boy, her little boy.

Concerning intellect, Rankin again failed to elevate his personal status with his talents. As an ideal student, he always completed his homework with relative ease. He performed well across all subjects having only faltered slightly in seventh-grade biology because dissections made him a little queasy. Still, being smart wasn't something your classmates automatically regarded favorably. In fact, if you weren't careful you could be donned with the dreaded nerd label. Rankin wasn't careful.

However, being smart and accepted as cool wasn't a hopeless endeavor. Borias pulled it off beautifully. He, too, ranked high in the brains department, but unlike Rankin he applied his wit to establish social acceptance. Rather than goofing off and baiting people to laugh at him, Borias preyed upon the awkwardness of others and stung them with clever insults. His barbs could be outright funny, borderline offensive, or on occasion, downright mean. Most of his classmates shied away from him because if he chose you to be the butt of his joke, you could end up looking pretty bad.

So, it was a tale of two brains. In one corner, Rankin hopped around like a bunny wearing high-water pants and devoted his attention to algebra. Knowing that x^2-1 could be successfully factored into $(x+1)(x-1)$ would someday help land him a swell salary working in a technical field, but for now it was just another arrow to place in his *please-kick-my-ass* quiver. Meanwhile, Borias landed knockout punches by rewording Handel's *Messiah* to broadcast the unfortunate hygienic condition of an elderly social-studies spinster.

Ha li to sis
Ha li to sis
Hali tosis
Hali tosis
Hallelujah

Rankin's social misery began in earnest last autumn at the start of eighth grade. Some regimented code of conduct infected his classmates and changed them into behavioral zealots. His peers no longer tolerated anything outside the mainstream. It wasn't just the usual critics. The majority of eighth and ninth graders at Collins Junior High seemed to be judging all aspects of your life, be it your clothes, friends, personal interests, or manner of speech. Rankin was doing nothing differently than he had the last couple of years, and that was exactly the problem. He hadn't grown up.

Why should he? Rankin was happy being himself. He had

never stood out as the class leader or the star athlete, but he was okay with that. Up until eighth grade, he played the role of a normal kid, and he enjoyed most of the things kids enjoy. He liked watching cartoons and eating candy. Boy, did he like eating candy. However, Rankin was now being judged by the treats he bought. While Candy Necklaces, Wacky Wafers, and Hot Tamales were acceptable last year, they were now regarded as something only little kids ate. Reese's Peanut Butter Cups were hip, but Bottle Caps were lame. Soda pop fell under the same scrutiny. Pepsi and Coke were righteous, but get caught drinking Grape Crush and you were going to catch some grief.

Cool teenagers watched cartoons like *Road Runner* or *Hong Kong Phooey*. Whenever Rankin's classmates learned he was still tuning into reruns of *Lidsville* and *H.R. Pupnstuff*, they would mock him in their imitation baby voices. Finally, while most of the guys discussed last night's episode of *Good Times* or *M*A*S*H*, Rankin happily recalled *The Wonderful World of Disney's* presentation of *Charlie, the Lonesome Cougar*. Because Rankin was not yet ready to exit from his happy place in childhood, he was ridiculed.

Rankin's size also worked against him. He had never been considered tall or husky, but for the past year his growth hormones parked in neutral while most of the class zoomed ahead. His diminutive stature made him easy pickin's for others looking to bolster their own ego. Ironically, most of the culprits harassing Rankin were other boys similar in size, and no one was guiltier than Virgil Weems.

One November day in woodshop, Rankin was being a little too meticulous while using the jigsaw to cut out his name from a plank of wood. His deliberate pace created a backup for the power tool.

"Quit beating off and finish your damn cut," Virgil yelled. If you're gonna spank your monkey, go do it the corner. Geez." Then Virgil uttered the term that would stick to the young Bellamy boy for the remainder of the year.

"Spankin' Rankin," Virgil muttered. He hadn't put any thought into it; it just flung out of his mouth.

"Spankin' Rankin!" Larry Clark echoed and then began howling in laughter.

Spankin' Rankin traveled through the school like wildfire, one of those uncontrollable incidents in a person's life having enormous implications. Rankin's identity had been completely transformed. Not a single school day passed in which the cry, Spankin' Rankin, didn't bounce off the hallway walls between classes.

At first, Rankin tried to laugh off the insult. Then he tried to ignore it, but the term endured and actually increased in usage over time. In fact, his unwanted title qualified as one of the most memorable items for the eighth grade class of 1975. Most girls in Rankin's class felt bad for their classmate and refrained from using the slur; however, even they couldn't help but snicker when the name came up in Rankin's absence.

Rankin, who had always been friendly, withdrew from his classmates and became less friendly. His new moniker brought added scrutiny to his affection for The Partridge Family, Red Hots, and substandard fashion attire. The nickname drew attention to his personal habits, and his behaviors reinforced the title forming a vicious cycle.

As 1974 gave way to 1975, the buzzards circled around Rankin's deflated carcass and grew bolder. For many of the bullies seeking dominance, verbal abuse wouldn't suffice. Rankin suffered physical harassment in the form of shoves against the lockers or books being knocked out of his hands. The tormenting culminated on the first Saturday in May.

Rankin had agreed to help Regan and some of her friends while they set up for the spring musical at the high school. The mostly grunt work entailed hauling miscellaneous equipment and taping sheet paper onto walls. When they finished, the group drove to the Pioneer Shopping Center where Regan and her friends entered the A&W. Rankin peeled off to visit Johnny's Hobbies and then Pioneer Drugstore with its amazing selection of candy costing a quarter or less.

Rather than hang around and listen to the girls yak about who's going with whom, Rankin decided to bask in the unusually warm spring day and walk home. Packed with thirty inches of red-rope licorice, a staple that hung on a bedroom hook when it wasn't being consumed, and Sugar Babies, he chose the most direct route leading him through a large expanse of vacant property. One end of the useless land abutted the junior high school. The area was splotched with whitish deposits of alkali soil, and intermittent flows of wastewater wound through a ravine. Some hardened dirt paths, where kids rode their bikes, ran across the terrain, and Rankin chose one that led toward his home.

About midway through the desolate area, the path descended into a streambed and turned sharply to the left hugging the steep hillside. As Rankin rounded the turn, he almost walked into Billy Mendoza. Billy was caught off guard and stepped back, but he regained his composure once he recognized Rankin.

"Goddamn, Rankin. What are you doing creepin' around here?" he demanded.

Billy's question drew the attention of two other boys who had been crouched over the water twenty yards behind Billy. Lance Chitwood and Virgil trotted to where Billy and Rankin were standing. The trio had decided to spend their restless Saturday afternoon moseying down to the badlands as they called it and having a smoke. As Virgil passed a pack of cigarettes to Billy, he interrogated Rankin.

"What are you doing here, Spankin' Rankin?"

Rankin grew defensive. "I'm just walking home," he said uneasily.

Virgil continued, "Ah, I don't believe that. You came here for a reason. Is this where you come to choke your chicken?"

"No," Rankin snapped.

"You're lying," Virgil accused. "I bet this little hideaway is where Rankin does his spankin'." Virgil smiled at Lance and Billy, and they smirked back. Virgil continued, "But I bet it's not your only place. Have you ever been caught beating off in your closet?"

"No."

"Then that's a good spot, too," Virgil said glancing at his two cohorts for approval.

Rankin tried a different tactic. "That's a good one, Virgil. You got me," he responded with a surrendering laugh.

"You think that's funny?" Virgil asked in a hostile tone. "I think you're a sick bastard."

"Everyone knows he's a fag," Lance chimed in.

"You know, if we get caught at school lightin' up, we get hauled down to the office and either get suspended, or Andersen gives us a swat." Virgil's mood had darkened. He raised his voice and spoke faster. "But they don't do anything to you, and you're a perv. You just get to skate free 'cause you're this little pussy teacher's pet."

"Maybe *he* needs a spanky," Billy suggested.

"Yeah, a spanky," Lance agreed. "Spanky Ranky."

Rankin panicked. "I need to go," and turned to leave the way he'd come.

Lance grabbed the bag of candy.

"Give it back," Rankin demanded as he reached to retake the bag.

Lance tossed the bag to Virgil.

"Give it to me, Virgil."

Virgil stared at Rankin. "Make me. You wanna fight? I'll kick your ass." Virgil and Rankin measured about the same size but stood light years apart in confidence.

Rankin quickly backed down and would have surrendered the candy in exchange for his freedom, but Billy had other thoughts.

"I think it's Spanky-Ranky time. He nodded to Lance, the largest in the group. Lance grabbed Rankin from behind and held him tightly.

"Grab his legs, and we'll put him on the ground," Lance directed Billy. Billy bent over to grasp Rankin around his calves.

"Stop it," Rankin cried and flailed wildly with his legs. His knee caught Billy in the mouth.

"Ow! Shit!" Billy yelled as he reeled backward.

Virgil stepped forward and slapped Rankin on the side of the head. Billy stood up and walloped the outnumbered boy with a hard punch into his stomach. The blow doubled Rankin over and he went limp in Lance's arms. Lance let him go, and Rankin fell to his knees.

"Time to whip your ass," Virgil screamed. "Lay him flat."

Billy pushed Rankin onto his belly. Without unbuttoning Rankin's jeans, Virgil pulled the boy's pants and underwear down around his knees. Virgil's mouth frothed as he swatted Rankin's butt with an open hand. After several strikes, Rankin's butt cheeks turned a bright crimson. The torture would probably have continued in both duration and severity had it not been for unexpected visitors.

A boy and girl, possibly brother and sister, of grade school age rode down around the path on their bikes and stopped abruptly when they saw the cluster of boys. Their arrival startled Lance, first, and then Billy. Finally, Virgil became aware of their presence.

"What are you lookin' at?" Billy yelled. "Get out of here!"

The riders didn't have to be told twice. They pushed past the older boys with their feet and started pedaling furiously, but the distraction they caused was enough to save Rankin. He bawled loudly. Virgil continued to breath heavily, but the rage on his face disappeared.

Whether they realized their crime had been witnessed or their own consciences kicked in, the gang's mood changed.

They looked back and forth at each other in silence. Finally, Virgil broke the ice.

"Let's get outta here."

The assailants rose. Billy lifted Rankin's bag of candy and threw into the water. The trio ran down the trail and out of sight.

Gradually, Rankin pulled up his underwear and jeans. He sat up and wallowed in a mixture of pain, fury, and shame for the next ten minutes. During his slow walk home, he wavered back and forth about telling his folks. However, his puffy, red eyes and broken spirit revealed everything to his dad when he crossed through the front door.

Martin Bellamy normally functioned in a quiet and thoughtful manner. When he wasn't teaching economics at the local junior college, he was engaged in tracking the Dow Jones Industrial Average or reading the latest Michener novel. Martin had some awareness of the difficulties Rankin endured at school—he had experienced similar episodes during his youth—but this brazen attack on his boy enraged him. Martin acted immediately and drove his Fiat to the Chitwood residence. Martin wasn't exactly chummy with Bill Chitwood, but they were acquaintances since the boys' days in the Cub Scouts.

When Bill confronted Lance about the incident in Martin's presence, the boy denied it outright. Unfortunately for Lance, when only two of the three guilty parties participate in a cover up, the facade of innocence crumbles. Teenage boys and long-term planning are mutually exclusive. When the trio parted ways earlier, they hadn't discussed how to handle the fallout. By his very nature, Virgil always denied any wrongdoing, so he wasn't a problem. Billy, on the other hand, couldn't keep his yap shut. He spilled the beans about Spanky Ranky to Larry Clark. Larry in turn shared with many. By nightfall, the story was well publicized, and the perpetrators had confessed.

Meanwhile, Mrs. Bellamy was livid back at the Bellamy home. She hated this hick town and reminded Martin, often, it had been a mistake to move here seven years ago. This episode only solidified her opinion. She wanted to press charges against the three boys, but Martin hadn't made up his mind. Bill Chitwood and Yolanda Mendoza, Billy's mother and a single mom, pleaded with the Bellamys not to get the police involved. They would see to it their sons apologized to Rankin and would be disciplined at home. Virgil's dad wasn't home when Martin stopped by, so he left a message with Virgil's older sister to have him call when he returned. Duggan Weems never did.

The Bellamys argued bitterly over a course of action. In the end, Martin won out by convincing his wife that avoiding legal action was in Rankin's best interest.

"Connie," Mr. Bellamy started, "Rankin's had a terrible day, but if we press charges then he's going to have to go to the police station and then court to testify. He's going to have to relive this day over and over. For his sake, I think we all just need to move on."

Mrs. Bellamy reluctantly acquiesced but not before receiving two promises from Martin. First, if any of these boys bothered Rankin again, they would be arrested. Second, Martin would go to school first thing on Monday to make Principal Andersen aware of the situation. Mr. Bellamy complied.

On Monday, the timing of the morning announcements

was delayed. At 7:55, Mr. Bellamy met with the principal. At 8:25, Virgil Weems, Lance Chitwood, and Billy Mendoza were summoned to Principal Andersen's office and immediately suspended. Since he had only been in school a total of twenty minutes, just a few people saw Virgil, but those who did witnessed a swollen black eye. Apparently, Virgil's sister had passed the message onto her dad. At 8:37, Mr. Andersen came on the loudspeaker.

"I am very disappointed to report we had an incident over the weekend where one of our students was assaulted by some other students. Even though the incident was not on school property or during school hours, the guilty individuals have been suspended from school in agreement with their parents. Apparently, the student who was attacked has been bullied here at school for some time. This behavior is unacceptable. Effective immediately, anyone caught demeaning a fellow student with derogatory slurs will be suspended, and anyone threatening or harassing this particular individual will be turned over to the police."

By now, virtually everyone knew the particular individual was Rankin. In his class, all eyes focused on him. He stared down at his desk steeped in embarrassment. Later that day, an inconsiderate ninth grader let out a *Spankin'-Rankin* call. Coach Dunkirk grabbed the boy by the back of the neck and forcibly escorted him to the office. During the afternoon announcements, Principal Andersen relayed the episode and noted the student was now under questioning by the police. The administration was dropping the hammer, and Rankin, once again, felt isolated in shame. During the final three weeks of school, nobody messed with Rankin. Hell, hardly anyone even talked to him. He was a pariah.

Summer vacation brought reprieve to Rankin's misery. He flew to Virginia with his mom and Regan for a visit with Mrs. Bellamy's folks. Rankin loved Granddad and Grandma Higgins. While Mrs. Bellamy and Regan's trip ended after two weeks, Rankin stayed on for another month. In light of what Rankin had endured, Martin and Constance agreed some time away from Casper would be good for the boy.

When Rankin returned, his actual resumption of Casper life wasn't nearly as bad as he dreaded. Time healed some of the sting. Given a chance to recover, the resiliency of his good nature reemerged, and he anticipated starting his first job. Well, it wasn't a real job, but he would earn some money. Mrs. DuPree hired Rankin to teach her twin daughters to play the piano.

The DuPrees had moved in across the street just over a year ago. For Rankin, their arrival counted as one of the few good things during an otherwise disastrous year.

Rankin and Jimbo hadn't shared any classes during eighth grade. Jimbo usually hung out with Randy and some of the other jocks but seemed like a good guy. Upon arrival to his new school, Jimbo swooped in as an immediate sensation with the girls; he was Casper's version of David Cassidy.

With adorable faces and bubbly dispositions, the twins, Eva and Eileen, dazzled new acquaintances as well. Fortunately for anyone outside the immediate family who wanted to distinguish between the two, they were born fraternal and not identical twins. However, they demonstrated solidarity in fashion as they dressed in matching outfits daily.

Mrs. DuPree radiated genuine beauty. Rankin doubted fair Helen of Troy possessed a more alluring face. The aging process added some roundness to a figure once as stunning as her smile, but Rankin didn't notice. He loved her. It wasn't lust but instead a deep, sentimental admiration. Mrs. DuPree displayed all the charm for which the South is renowned and did so in a kind and unassuming manner. Her delicate accent soothed Rankin.

To top it off, Mrs. DuPree had mastered the art of cooking, and homemade desserts were her specialty. During Christmas season last winter, she baked holiday trays for the Bellamys and the other families on the block. She shared a delicious variety of divinity in shades of both pink and white, tiny pecan pies, and the best fudge Rankin had ever tasted. However, Laurie DuPree's pralines rose above all the other yummies. When Rankin bit into the creamy candy for the first time, his taste buds rejoiced.

Rankin knew Eva and Eileen fairly well. He often waved to them when he returned from school, and once he babysat them. Normally, Regan did the babysitting, but she had forgotten about a school engagement when she committed to cover them one evening. Rankin and Mrs. Bellamy rescued her. Because Rankin carried rookie status at sitting, the twins came over to the Bellamy house where his mom could assist. Rankin introduced the little girls to the piano and entertained them with some silly songs.

The notion to hire Rankin as a piano teacher dawned on Mrs. DuPree when Bo and she picked up the girls. After the Spanky Ranky incident, an event now known to half the town, she wanted to help the boy in some way and saw the piano lessons as a chance to cut Rankin a break. Mrs. DuPree didn't anger often, but like Mrs. Bellamy, she grew enraged when she heard what happened. She administered the third degree on Jimbo to determine if he had any involvement. He hadn't but did admit to thinking Spankin' Rankin was funny when he first heard it. He had actually mentioned the name to Mr. DuPree when it first occurred and recalled his dad cracking up as well. However, when he glanced toward his dad for some moral support, Mr. DuPree wore a solemn expression on his face. Bo DuPree knew his wife and when to keep his mouth shut.

Rankin thoroughly enjoyed his role as piano instructor. He wished the girls were more serious, but as Regan reminded him, they were only five. Eva indicated a greater interest and accordingly was becoming the better player. The lessons began in mid-July, and by the second week in August Rankin calculated he had heard some form of *Twinkle Twinkle Little Star* at least one hundred times. Because summer vacation afforded him plenty of free time, Rankin met with the girls twice a week.

On a lazy August afternoon, Rankin promised the twins a special surprise if they concentrated on their lesson. They complied, and Rankin brought out frozen Push Ups. Eileen chose grape while Eva and Rankin picked cherry. Mrs. Bellamy, who wanted no part of sticky drips in her kitchen, shooed them outside to eat their treats. They marched into the backyard and sat down in the shade of a cottonwood tree. For a few minutes, quiet prevailed as each kid licked down the ice cream. Then Eva broke the silence.

"What does Spankin' Rankin mean?" she asked innocently.

The memories of the past school year rushed back into

Rankin's head. Although miscellaneous reminders of his unwanted nickname still emerged, enough time had passed that he might go for hours without thinking about this unpleasant matter. Now out of the blue, it infiltrated his deepest sanctuary. Rankin tried to avoid the subject by grabbing Eva and tickling her. The sweet little girl with ringlets of blonde hair squealed and wriggled until she escaped.

As Eva stepped away from Rankin's reach, Eileen plopped herself in the boy's lap. She stood on his folded legs and put her face with those big blue eyes and newly acquired grape moustache directly in front of his.

"What does Spankin' Rankin mean?" she asked repeating her sister's question.

The passage into adulthood takes many forms. Physical changes are easily noticed. Girls begin to menstruate and boys' voices deepen. On the emotional front, individuals start taking responsibility for their own behavior. A guy confesses the dog didn't eat the homework and admits he made a conscious choice not to do the assignment. Subtle changes also occur. An innocently optimistic outlook on the world fades and makes room for the introduction of cynicism. For the first time in his life, Rankin was about to exhibit this latter condition.

"What does Spankin' Rankin mean?" Eileen again pressed.

"It means even dumb asses can rhyme," Rankin finally replied.

"Umber! You just cursed," cried Eileen in shocked surprise. Still deep into childhood, she locked onto the words and missed the message entirely.

A few weeks later, Rankin sat back and assessed the beginning of ninth grade. To date, he counted mixed results. On the negative side, he rubbed the red and now scabbing abrasion on his face complements of Totem. What should have been a huge self-confidence surge turned into a beat down when Rankin eliminated Totem from a game of dodgeball. The big Indian went berserk and attacked Rankin. Totem was suspended from school, and Rankin was left wondering how playing a game well still resulted in getting his butt kicked. Still, the gym incident felt different. Instead of mean-spirited teasing, Totem's anger seemed spontaneous.

On the bright side, the Spankin' Rankin craze seemed to have passed. On the first day of school, Rankin walked by Virgil and Lance in the cafetorium. No words were exchanged. He had heard Billy's mother remarried, and the family moved to Colorado. At least for now, anyway, no one else was interested in riding his case.

With *Road Runner* playing on the TV, Rankin tossed back another swig from his Pepsi and thought some more. He felt different this year. He had acquired a greater interest in the clothes he wore, and he was starting to notice girls. Wow, some of them had really changed over the summer, especially Helen Barrow. Swellin' Helen! *Did he just say that?*

THE BOYS

"Wow, Totem," Jerry teased. "You're up to fifty-eight miles per hour. Don't over do it now."

"Hey, man, my luck hasn't been so great lately. I'm already three miles over the speed limit," Totem responded. "I'm not gonna push it. I don't need a ticket."

"Choate says you can go up to sixty-four, and no one will bother you," Jerry advised from his position in the front passenger seat.

"Who's Choate?" Jimbo asked from the back of the car.

"My mom's squeeze," Jerry replied. Bobby Choate had married Jerry's mom last February technically making him Jerry's step dad, but Jerry would never refer to him as a dad in any way—ever.

The three boys rode in a white Dodge Polara with blue interior on the old Yellowstone Highway between Glenrock and Casper. They were feeling the slow-motion effects from newly-enacted legislation regarding the oil crisis. In 1974, the Emergency Highway Energy Conservation Act was executed limiting the maximum speed on highways to fifty-five mph.

The law was a response to the consequences of a war that

took place halfway round the world. In October of 1973, the Yom Kippur War broke out between Arabs and Jews in the Middle East. These enemies battled for the fourth time since the Jewish state of Israel had been formed in 1948. The three-week conflict began when Egypt and Syria launched a coordinated strike against Israel. The Arab countries surprised their enemy by attacking during the holiest day of the Jewish calendar and advanced deep into Israeli territory during the first few days.

Although the direct fighting occurred in a tiny region of the world, it represented a piece of the Cold War, the ongoing, tense standoff between the Soviet Union and the United States, the world's two superpowers. The Soviets backed Egypt and Syria with equipment; when Israel's supplies and munitions quickly depleted, the U.S. stepped in and replenished them. The U.S. aid turned the tide of the battle and enabled Israel to regain its lost ground. On October 26, 1973, the United Nations negotiated a cease-fire agreement.

To punish the U.S. and other countries for supporting Israel, the Organization of Petroleum Exporting Countries (OPEC), a group of primarily Arab nations, imposed an oil embargo. The embargo substantially cut oil flow to the U.S. at a time when domestic oil production had declined. The price of oil jumped from under three dollars per barrel to almost twelve dollars in a matter of months. Combined with a stock market crash a few months earlier, the American economy lie crippled in a state of shock. As part of a national effort to conserve energy, the speed limit on highways across the country was reduced.

The effect of the speed limit reduction was notable. The change hit drivers in western states who had to traverse large expanses of land especially hard. An extra two hours was required to complete a trip from Casper to Denver. To the chagrin of drivers everywhere, vroom vroom had been reduced to putt putt.

The boys were returning from a babe-prospecting mission. Having scored a phone number from a Glenrock High School cheerleader a few weeks back, Jimbo placed a call earlier in the week and arranged to visit. The guys decided to truck on over to the little town during their free afternoon on Friday. Students had been dismissed at noon throughout the Casper School District to accommodate an in-service teachers' day.

The guys planned to meet Patsy, Terri, and a girl to be named later at 2:30 in the parking lot of Happy's Drive Thru. According to Patsy, Happy's claimed the title as the premier, teen, hang-out spot in downtown Glenrock, an urban mecca consisting of two blocks. Although the plan sounded good, its execution disappointed the hopeful boys to say the least. Patsy, Jimbo's personal interest, was a no show. She failed to mention during their telephone conversation she had just started dating one of her goat-roping classmates. Sometime between Tuesday's phone call and today, Jimbo's competition had planned a different adventure that afternoon for Patsy. Next, the girl to be named later and Jerry's potential match arrived in the form of Terri's thirteen-year-old cousin, Dink. Really? Jerry refused to babysit an undeveloped and, quite possibly, inbred tot.

At least Terri's appearance partially redeemed this sad rendezvous. She was pretty cool and exercised her status as a part-time employee of Happy's to land the boys free drinks. Terri told them she couldn't stay long as she had to get ready to cheer for a home football game, but if they wanted to come back down tomorrow night she'd be working behind the counter until 11:00 and could hang out afterward.

Jimbo and she spoke most of the time, but Totem studied her with interest. Unlike their first meeting when she wore a baseball cap, Totem observed how her cinnamon-colored hair carried a bit of a curl and how it contrasted favorably with her deep brown eyes. He also noticed a small scar above her right eye partially hidden by her eyebrow. It looked like an old wound, and he guessed it was a remnant from some childhood accident, probably one of those situations where you're recklessly zipping around the house and catch the corner of the kitchen counter with your head.

Totem also sensed a mischievous wryness to her personality. She acted carefree and fun. He thought she was a girl who might get him to relax and open up, a behavioral trait difficult for him.

On their way to Glenrock, the optimistic threesome drove on I-25, the efficient choice. However, because the gathering ended much sooner than they expected, they owned some free time and decided to dawdle back to Casper via the old highway. The two-lane road still accommodated local traffic and sporadic visits by disoriented tourists.

"Well, that was a wasted effort," Jerry summarized succinctly.

"I eee," Jimbo seconded.

Still, the guys were upbeat. Hey, they were sharing togetherness on a weekday afternoon free from school. They cruised down the road in a car by themselves, no parents. In particular, Totem was enjoying himself more than he had in a long time. He didn't think the trip had been a wasted effort. Plus, he was driving. He loved goosing the accelerator on a hill and bending the car around tight curves. Behind the wheel, he fully controlled the situation. For Totem and his riders on this day, living felt grand.

The terrain between Glenrock and Casper included a roller-coaster series of climbs and descents. Upon cresting one of the larger hills, an impressive panorama graced the teenager's eyes. To his right, Totem viewed the North Platte River with yellow-leaved cottonwood trees adorning its banks. In front of him and slightly to the left, he scanned the unhindered expanse of Casper Mountain's east side. Looking further west, he framed Casper Mountain Road in the center of his windshield. The road carved into the mountain's side and provided a clearly visible human landmark.

As they drew closer to Casper, an unusual rock outcropping hosting a small forest of pine trees appeared on the car's right side. Totem had hiked this familiar area many times with his father and brothers when he was younger. Along the outcropping's edge and about five hundred feet from the road, both Totem and Jerry noticed three figures standing apart from one another. Each appeared to be looking down as if they were searching for something. Jerry was first to recognize one of them.

"Hey! Is that Winters?" he said in a half questioning, half stipulating tone. "Slow down."

Totem complied. When Jimbo confirmed it was, indeed, Randy, the driver pulled off to the side of the road.

"Holy Hell! Who else is with him?" Jerry wondered aloud.

Everyone in the car strained their eyes to identify the other guys standing in the open field. For their part, the guys in the field hadn't initially taken notice of the vehicle, even when it pulled over and stopped.

"That's Borias," Jerry announced as he labeled the second body.

"And the other one looks like Rankin," Jimbo added.

Totem's stomach churned when he heard the last name— bad memory. He noticed each person was holding something. They looked like medium-size canvas bags.

"What are they doing?" Jimbo asked.

"Don't know," Jerry replied. He rolled down his window and then cupped his hands around his mouth in megaphone fashion. "Hey, Winters!"

When Randy heard Jerry's voice, he lifted his head and began trotting toward the car. Once Randy committed, Borias and Rankin began walking toward the car, too. As Randy neared the car, he turned back to the stragglers and called out, "It's Jerry and Totem."

Borias continued moving forward, but Rankin stopped dead in his tracks.

"What're you doin' out here?" Jerry asked.

Randy stepped five more paces to reach the passenger side of the Dodge. Then he said, "We're huntin' for rocks. We've got to turn in a rock collection for science next Monday." Then he turned back toward Rankin and Borias and yelled, "Jimbo's here too." He faced the car again and added, "Hey, Jimbo."

"I eee," Jimbo acknowledged.

Jerry began dogging him. "Your project is due in three days, and you're just now getting started?"

"We've been working on it," Randy fired back. "And I said next Monday. Do you have shit in your ears?" If Jerry wanted to tangle, Randy was ready. He relished the smack talk. Then he became the inquisitor. "Where have you guys been?"

"We've been on a pussy quest," Jerry boasted with satisfaction.

In a purposely drawn out manner, Randy shielded his face with both hands and peered into both the front and back seats. "Mighty successful, I see. Excellent work, men," he said grinning and making the A-Okay sign with his hand. "Of course, you have the golden boy sitting in the back seat instead of the front, so that explains a lot."

By this time, Borias had reached the car. "Hey, man," he said to the group as a whole. He looked the car over and then peeked in at Totem. "Wow! Nice car. Is it yours?"

Totem nodded but didn't say anything. Randy looked at Totem and decided it was time to give the driver his due. As he walked around the back of the car to get to the driver's side, he considered the Indian carefully.

Randy savored exchanging burns and getting a rise out of people. He was good at it, so good in fact it had often become too easy. Take Rankin for instance. If challenged, he'd just roll on his back to expose his belly in surrender. Jerry would put up a fight and might even land a few barbs, but in the end, Randy would win. He avoided matching wits too often with Borias because Borias was a deadly counter puncher. Fortunately for Randy, Borias rarely engaged in the back and forth game, but instead preferred to assail unsuspecting victims like a sniper using a single precise kill shot.

Then there was Totem, the crown jewel for Randy. Totem's verbal talent didn't qualify him for such high regard. Are you kidding? The guy hardly spoke. Instead, Totem's allure lie in his stoic nature. He was as stone-faced and expressionless as they came. Randy savored those occasions when he could elicit an uncomfortable reaction from Totem.

Randy checked to make sure there were no oncoming cars. Then he stood next to Totem's door, raised his right hand, and with the most serious face he could muster, uttered his greeting.

"How, Chief."

Having dealt with this routine so many times before, Totem was fully prepared. He responded by rubbing the bridge of his nose with an extended middle finger.

Undeterred, Randy continued by pointing to the west. "You, take 'em scouting party back to big fort."

Totem grabbed his drink cup from Happy's, now containing only water melted from ice cubes. He sipped a mouthful with his straw. Then using his tongue and teeth, he shot an arched stream of water toward Randy, hitting his adversary on the side of his face.

Randy staggered backward in surprise. "Ah! You bitch!"

Totem almost smiled.

"Cute baggy ya got there," added Jerry piling on.

The reference to the rock bag reminded Borias about his

own collection. He reached into his sack and retrieved a piece of whittled flint.

"I found an arrowhead!" he grinned and exclaimed. This outing afforded a new experience for Borias. He'd never before participated in a rock-hunting excursion and was thrilled to have spotted this unexpected treasure.

In response, Randy dug out a dull stone from his bag and made a similar claim. He held it up for Totem to inspect.

"Totem," Randy began, "I think this is yours. Were you hunting rabbits or ugly women?"

"I thought I saw your fat ass, but it turned out to be a cow."

"A cow!" Randy exclaimed. "Jerry wasn't with you, was he? 'Cause if he was, he'd try to hump it. Probably take control of the wheel, run off the road, chase the poor animal down, and hump, hump, hump."

"Shut it off!" Jerry barked.

Borias interrupted the fray. "You think you can give us a ride back to town?"

"Of course, they can give us a ride," Randy assured him. "But it does mean they'll have to say goodbye to their lady friends." As if he were playing charades, Randy opened the back seat door and motioned for the invisible women to exit. "Okay, girls, out of the car. Scram!"

Jimbo doubled up in laughter at Randy's mockery. Then he redirected the group's attention. "Why's Rankin just standing out there?" he asked.

"That little goof," Jerry added. Totem remained silent.

God, do I have to take care of everything? Randy wondered

silently. Aloud, he said, "I'll go get him. You boys don't touch any buttons 'til I get back."

Rankin had been standing by himself in the open field since hearing Totem's name, watching from afar. He set down his bag of rocks and put his hands in his pockets, but he hadn't taken a step closer to the car. He looked on as Randy marched toward him.

Using an awful John Wayne impression, Randy addressed the lone boy. "Saddle up there, Pilgrim. We've got ourselves a ride into town."

"Um, no," Rankin mumbled. "You guys go on. I'm just gonna walk to the truck stop like we planned."

"No can do Little Britches," Randy answered. "If I leave you out here alone, you'll get eaten by a prairie dog, and then your mom will beat the hell out of me. Or worse yet, she'll make me go to one of your sister's concerts. *Ha!*"

Randy drew a smile from Rankin with the last comment. They had lived on the same street for a long time and knew the ins and outs of each other's families. Rankin's smile didn't last long.

"I don't think so," he repeated.

Randy's demeanor turned serious. "C'mon. It'll be cool. I won't let anything happen."

Rankin still had reservations, but between his wishy washiness and Randy's persuasion he found himself grudgingly walking toward the car. He thought about prairie dogs for a moment. He knew they didn't eat people—that was silly—but they were known to carry rabies.

"Prairie dogs don't really attack people, do they?" he asked.

Randy answered by putting him in a loving headlock as they headed toward the others.

"Move out of there, Wimpy," Randy directed Jerry as he reached the car. "Let the master have shotgun."

"*You* are in the back, Jack," came Jerry's reply.

Randy knew Jerry wouldn't give up the front seat, but it was worth a try. Jerry stepped out of the car only long enough to let Borias move into the middle of the front bench seat. Jimbo slid over behind Totem, and Rankin then Randy hopped into the back. Totem glanced into the rearview mirror briefly making eye contact with Rankin before each quickly looked away.

Totem waited for a pickup and a sedan to go by before pulling back onto the road. As the car accelerated, Jimbo began quizzing the newcomers about their field trip.

"How did you guys get out here?"

"It's a secret. I don't want to have to kill you," Randy replied.

"Aw, let 'em in on it," added Borias as he played along. "We walked into the transporter room and told Scotty to beam us out to some shit-hole pasture." Jerry and Jimbo burst into laughter.

"Mrs. Borias gave us a ride," confided Rankin, completely killing the moment.

"Was your mom okay with you coming out here?" Jimbo asked Rankin. Because he also lived on the same street as Rankin and Randy, Jimbo knew about Mrs. Bellamy and her protective habits.

"Well, I didn't exactly tell her. I just said Mrs. Borias was taking us to collect some rocks for our science project," Rankin confessed.

"Rankin, you sneaky little bastard," said Jerry.

"Good job!" added Jimbo. "Give me some skin," he said and offered up his palm.

Surprised, Rankin tentatively slid his hand across Jimbo's. Maybe he would get out of this car alive, after all.

Randy watched his friends' interactions silently. He had been concentrating on Totem and plotting his next assault, but he noticed the driver was clearly uncomfortable having Rankin in the car. He decided it was time to clear the air.

"Ahem," Randy began by clearing his throat. "Before we get any further down the road," he started and then paused looking to see if everyone was paying attention. "That's a clever play on words, see, 'cause we're actually moving down the road."

The gang responded severely.

"Shut up."

"Oh Gawd."

"Give it a rest."

Randy continued past the groans. "We need to fix a little problem between two of you."

Totem's back stiffened, and Rankin's eyes grew large. The altercation had taken place almost a month ago. Rankin instinctively rubbed his face to feel the scab, but it was now gone.

"A few weeks ago there was a little spat, and since we're all *pal-ies* now it's time to kiss and make up."

Jimbo leaned forward in his seat anxious to see what Randy was cooking up. Jerry turned around and stared at Randy with an *are-you-crazy* look. Rankin felt a grapefruit growing in his throat, Totem squeezed the steering wheel, and Borias nonchalantly kept nodding his head. Like a judge issuing sentence, Randy commanded the situation.

"Totem, you're first."

Totem prepared to stop the car and smack Randy in the head.

"Repeat after me," Randy ordered. "Rankin, I'm sorry I smeared your face into the gym floor."

Borias snorted out a laugh while Totem glanced back at Randy.

"Say it," Randy insisted.

Totem wasn't sure why, but he reluctantly complied. "Rankin, I'm sorry I smeared your face into the gym floor." Jerry covered his smile with his hand.

"See, that wasn't so hard now, was it," Randy said flatly. "Okay, Rankin, now it's your turn to apologize."

"I'm sorry," Rankin blurted.

A look of disgust formed on Randy's face.

"Rankin, stop! You have to wait until I tell you what you're apologizing for. Geez."

Poor Rankin was completely flustered. He was just trying to get along.

"Okay, Rankin, are you ready?"

"Yeah."

"Are you sure?"

"Yeah."

"Repeat after me. Totem, I apologize for taking advantage of you. I know you're the shittiest dodge ball player on Earth, and I shouldn't have picked on you."

Hysterical hoots and shouts filled the cabin. Even Totem broke down and displayed a sheepish grin.

Rankin, now fully relieved, faithfully complied. "Totem, I apologize for taking advantage of you. I know you're the shittiest dodge ball player on Earth, and I shouldn't have picked on you."

Totem looked into his rearview mirror and again made eye contact with Rankin. This time both boys held the gaze long enough to silently acknowledge the matter was finished.

Jimbo sat back against the seat, stuck his arm out the window, and let the air pass through his spread fingers.

"Man, what a great day," he said, and it was.

In Wyoming, the weather on the third day in October can come in a variety of flavors. Snow is not uncommon, and winds can bluster, but today's temperature registered a sunny seventy-one degrees with just enough movement of air to tickle the hairs on your arm.

Jerry concurred with Jimbo's assessment and added, "This would be a great day for fishing."

Totem agreed with Jerry's comment. They had fished together many times. The passengers enjoyed a momentary silence with each boy caught up in his own thoughts. Then, Borias broke the silence.

"I've never been fishing before."

"Me, neither," said Rankin.

Totem was shocked. *Never been fishing? That's unheard of*, he thought. He felt sorry for these guys and was trying to imagine a life without fishing when Jimbo asked him a question.

"Hey, Totem, have you got your regular license yet?"

Totem looked at Jimbo in the rearview mirror and shook his head.

"Twelve more days," he said.

"And then you'll be driving all the time," Jimbo added aloud but really directed at himself more than anyone else in the car. "I eee," he crowed. Jimbo's call irritated Borias.

"I eee, I eee. What's that supposed to mean?" Borias asked turning his head to see Jimbo.

"It means I eee," smiled Jimbo.

Borias retaliated, "Why not eee I, eee I?"

"O," said a small voice. The one-syllable addition stunned both Borias and Jimbo.

"Good one, Rankin," complemented Randy after a brief delay. He had been caught off guard as well.

Rankin sat up straight sporting a stupid grin. *This is a new kind of attention*, he thought to himself. *I like it.*

The boys were entering the city limits when Borias threw out an unexpected proposition.

"Do you guys have somewhere to go right away?" The dashboard clock displayed 4:00.

Totem and Jerry looked at each other. After a moment, Jerry said, "Not really, why?"

"Let's drive around some more," Borias suggested. He was really enjoying himself and being surrounded by the guys.

"Driving takes gas, and gas costs money," Totem warned. He wasn't keen on spending his limited cash joy riding.

"Oh, I'll give you some gas money," Borias offered. He pulled a five-dollar bill out of his wallet and laid it on the dashboard.

Totem's eyes widened. He never remembered Jerry, Jimbo, or anyone else giving him gas money. "Okay," he said. He turned left at the first stoplight and drove two blocks where he caught the on ramp to the interstate highway.

Jimbo looked at Randy who shrugged with a *what-the-hell* kind of acceptance. Rankin chewed on the neckline of his red T-shirt, but this car ride had revealed many pleasant surprises. He, too, was willing to go with the flow. Now that the subject of plans had been raised, Randy threw out a new question.

"We were talking out in the field earlier about catching the Monty Python movie tonight. You fellers interested?" The question was addressed to the Glenrock trio.

The movie revelation jolted Jerry. Since when did Randy spend Friday nights hanging out with Borias and Rankin? "You mean Borias and you? And Rankin?" The words slipped out in a condescending manner before Jerry's common sense could filter them.

"That would be correct, Girard," answered Randy. His expression back to Jerry silently said, *Let's not be an asshole, shall we.*

Jerry's face flushed, but to his credit he corrected his mistake. Looking back at Rankin, he muttered, "Sorry, man. I didn't mean it like that."

Rankin had weathered much worse recently and appreciated the fact Jerry bothered to apologize. "No sweat," he said.

Moving the conversation back on topic, Randy asked, "What do you say, boys?"

Totem answered first. "I can't. Goin' deer hunting with

Father tomorrow morning. We're leaving at 4:00, so I gotta lay out my gear tonight."

"I can't either," Jimbo added.

"What're you doin'?" Randy pressed.

Jimbo didn't answer at first. He wanted to choose his words carefully. Finally, he said, "I told a friend I'd go roller skating."

Randy refused to let that answer go by unchecked—time to put on his Columbo trench coat. "Tell us about your friend," he coaxed.

"It's Inez," Jimbo revealed. He tried to contain a huge smile and failed miserably.

"Inez Ortiz!" Totem exclaimed.

"No way!" Randy shouted.

Borias jumped in. "You're going with Her High-nez?"

Even Rankin contributed a *that's-impressive* whistle.

"I'm not going with her. We're just going skating," Jimbo clarified.

Inez Ortiz, a.k.a. Her High-nez, was considered the most popular girl in their grade. She was the product of a Mediterranean marriage. Her Hispanic father operated a small construction business, and her mom's family was only two generations removed from Italy. Inez featured beautiful olive skin and brown hair so dark in color it could be mistaken for black. Yet, her mysterious eyes garnered acclaim as her signature trait. Their mesmerizing shade of green frequently bewitched not only the boys in her class but guys much older as well.

Her High-nez, one of the few nicknames not incubated in the mind of Borias, slighted her unfairly. Inez was as down to earth as they came, her only fault being she really had no faults.

For that crime, she drew the ire of other people's envy. A more apt description for her might be Mary Poppins. No, that didn't exactly work, either. Inez clearly had her act together, but practically perfect in every way wasn't how she came across. Too many differences could be drawn between Inez and the Julie Andrews character. Inez acted neither formal nor prissy, and it was highly doubtful half the ninth-grade boys at Collins daydreamed about boinking an English nanny.

"Is it just you two going skatin'?" Jerry inquired.

"I don't know. I'm meeting her at the rink," answered Jimbo. "Why? You wanna come?"

"I ain't goin' skatin'," Jerry stated emphatically.

Jimbo completed the discussion, "Whatever."

Seeking to refocus on the joy of the ride, Borias questioned Totem.

"You got any jams we can play?"

Borias had first spotted the under mounted 8-track player when he slid into the front seat. The player and four hi-fi speakers—two in front and two in back—had been installed by Edward a couple of years ago and counted as the single worthwhile enhancement Totem's brother had made to the car.

Jimbo echoed Borias' sentiment. "Yeah, play some *tune-age*."

Sitting shotgun, Jerry exercised the responsibility of opening the glove box and selecting a tape. He chose from four possibilities and picked one that still remained popular more than a year after its release. He waved his choice high in the air and received overwhelming approval from the back seat.

"BTO, BTO," Randy yelled as the sounds of Bachman Turner Overdrive II were hailed.

Jerry inserted the tape into the player. The channel one light illuminated and was soon followed by a single, low, sustained guitar note. The opening guitar riff ensued, and all the guys, even Rankin, were immersed in the groove. What happened next isn't the stuff of legends, but its rarity cannot be overstated. The six became one—at least in spirit, at least for an instant.

Jimbo again extended his hand out the window. The weather was perfect—not too hot, not too cold, not too windy. At this moment, he could temporarily admit he loved living in Casper, and he was certainly looking forward to an evening with Inez.

Sitting next to him, Rankin smiled with a happy radiance. Much had happened this afternoon. Not only had he avoided getting clobbered by Totem a second time, he felt as if a small connection had been forged with the car's driver. They weren't best buddies by any stretch of the imagination, but he now felt comfortable enough to see the tall boy with long black hair and ask, *How's it going?* He had also delivered the punch line to a joke, and the guys laughed. Finally, he'd established some independence even if it required some wording ambiguity and participated in an unauthorized adventure. He felt an adult type of satisfaction.

Speaking of satisfied, Randy glowed as he listened to his favorite band. He had engineered a peace settlement from the back seat and worked Totem into submission in the process. Today, he performed like a star; Zonk would be proud.

Likewise, Jerry felt at ease and in control. He rode shotgun. He selected the music. Best of all, he was hanging out with his good buddy just like old times. But these weren't old times;

these were new times, and they were new times with new friends. Randy, Jimbo, and Borias were good dudes and funny as hell. Heck, even Rankin was all right.

Borias sat back and let the song flow into his head. Today, he was introduced to an entirely new perspective. For once, he wasn't a loner on the outside looking in. For the first time in his life, he belonged to something greater. He loved the sense of camaraderie generated by squeezing six guys into a vehicle and letting them spout off about nothing and everything.

Behind the wheel, Totem enjoyed tremendous relief as the weight of the world lifted, at least a little. In his own irritating way, Randy had done him a huge favor in dealing with Rankin. The guilt had disappeared. Maybe the breeze had shifted. Maybe he wouldn't have to walk against the wind for a while.

Now safely back in his vessel and behind his own version of the wheel, PUUK observed the car and its passengers. For that one special instant, the traveler believed the boys were floating in some surreal, narcotic state. Inside the Dodge, the song finished.

If an old man is lucky, he will have to use his fingers and his toes to count the number of times in his life when he has experienced that perfect moment. The sensation lasts no longer than a breath, but while intact he realizes life simply can't get any better.

That sense of perfection is rare for any single person to encounter, but today, not one but six boys shared that incredible feeling simultaneously. How can you possibly explain this amazing coincidence to someone? Had they been aware of his presence, how could the boys have explained this moment to PUUK? It was, hmmm. It was *five sigma*! Yeah, baby, *five sigma*.

If you enjoyed reading MEET THE BOYS OF CASPER, stay tuned for the second book in the series, REMEMBER THE DAYS.